HOW TO SWAP
GM LT-SERIES
ENGINES INTO ALMOST ANYTHING

Jefferson Bryant

CarTech®

CarTech®

CarTech®, Inc.
838 Lake Street South
Forest Lake, MN 55025
Phone: 651-277-1200 or 800-551-4754
Fax: 651-277-1203
www.cartechbooks.com

Edit by Bob Wilson
Layout by Connie DeFlorin

ISBN 978-1-61325-388-5
Item No. SA411

Library of Congress Cataloging-in-Publication Data

Names: Bryant, Jefferson, author.
Title: Acumen : how to swap GM LT-series engines into almost anything / Jefferson Bryant.
Description: Forest Lake, MN : CarTech, Inc., 2020.
Identifiers: LCCN 2020011779 | ISBN 9781613253885 (paperback)
Subjects: LCSH: General Motors automobiles—Motors—Modification—Handbooks, manuals, etc. | General Motors automobiles—Parts—Handbooks, manuals, etc. | General Motors automobiles—Performance—Handbooks, manuals, etc.
Classification: LCC TL215.G4 B787 2020 | DDC 629.25/040288—dc23
LC record available at https://lccn.loc.gov/2020011779

Written, edited, and designed in the U.S.A.
Printed in China
10 9 8 7 6 5 4 3 2 1

DISTRIBUTION BY:

Europe
PGUK
63 Hatton Garden
London EC1N 8LE, England
Phone: 020 7061 1980 • Fax: 020 7242 3725
www.pguk.co.uk

Australia
Renniks Publications Ltd.
3/37-39 Green Street
Banksmeadow, NSW 2109, Australia
Phone: 2 9695 7055 • Fax: 2 9695 7355
www.renniks.com

Canada
Login Canada
300 Saulteaux Crescent
Winnipeg, MB, R3J 3T2 Canada
Phone: 800 665 1148 • Fax: 800 665 0103
www.lb.ca

CONTENTS

Acknowledgments 4
Introduction 4

Chapter 1: What Is the Gen V LT-Series Engine? 5
LT Car Engines 5
Gen V Truck Engines 6
Engine-Swap Projects 8
Swapping Problems 9

Chapter 2: Engine and Transmission Mounts...12
Motor Mounts 12
Adapter Plates 13
GM Truck Mounts 18
Transmission Mounts 20

Chapter 3: Oil Pans 26
Factory Oil Pans 26
Aftermarket Oil Pans 28
Oil System Modifications 38

Chapter 4: Accessory Drives and Cooling Systems 40
Stock Drives...40
AC Compressor Retrofit 44
Aftermarket Drives 47
Aftermarket Accessory Drive Assembly...........49
Cooling Systems 56
Gauge Sensors 59

Chapter 5: Transmissions 67
Automatic 67
Manual 74

Chapter 6: Wiring 86
Factory Harness......................................86
Aftermarket Harnesses 90
Sensors 93

Chapter 7: ECMs and Controllers 102
ECMs...102
Sensors 104
Throttle Pedal 106
Tuning...107
DOD Delete 111
Transmission Controllers.......................114

Chapter 8: Fuel System 115
Fuel Pumps..117
Fuel Control Module Pressure Sensor 122
Return Lines...126

Chapter 9: Exhaust Systems 133
Exhaust Manifolds..................................133
Aftermarket Headers...............................140
Catalytic Converters 142
Air Intake 143

Source Guide 144

ACKNOWLEDGMENTS

This book is representative of three years of intensive study and work on the subject of Gen V LT-series engine swaps. It has been a long road, and I relied on many people to help put this information together in one place to help you perform an LT swap with as few remaining questions as possible.

The following people were instrumental in accomplishing this goal: Matt Graves, American Powertrain; Curt Collins, Bill Nye, and Adrien Peters, Chevrolet Performance; Chris Douglas and Trent Goodwin, Comp Cams Group; Bill Tichenor, Blane Burnett, and Jeff Teel, Holley Performance; Jeff Lee, Martin & Co Advertising; Jeff Abbott Jr., Painless Performance; Don Lindfors, Pertronix; and Pat McElreath, Chris Franklin, and Bodie Hunt, Red Dirt Rodz.

Thanks to: Aeromotive Inc, Autometer, Current Performance, DeatschWerks, Dirty Dingo Motorsports, Dynotech Engineering, G Force Performance, HP Tuners, ICT Billet, LS1tech.com, Magnaflow, ProCharger, Street & Performance, Summit Racing, and Trans-Dapt.

Special thanks go out to my wife, Ammie, and my children, Jason, Josie, and Ben, for their patience and assistance in completing this project.

INTRODUCTION

The Gen V engine is the pinnacle of pushrod-engine technology and is capable of incredible performance due to its direct-injection fuel design. Since its release in 2014, the Gen V LT-series engine has proven to be a powerhouse, providing big horsepower with excellent fuel economy. There are several pitfalls that a swapper can run into when performing an LT engine swap, the main of which is the fuel system. Gen V engines require a pulse-width modulated (PWM) pump to accurately control the engine, and while swaps can be performed without the PWM system, the results are never as good as when the PWM is utilized.

Figuring out what is needed for a specific application can be difficult, but the goal of this book is to help provide those answers. Whether using a brand-new crate LT1 or a take-out 5.3 L83 from a salvage yard, this book will help you complete your project. Most swap vehicles require a power steering pump, but none of the Gen V series engines come with one. This situation is easily rectified with one of the many aftermarket options, including the new Chevrolet PS pump standalone add-on kit.

The aftermarket has now fully embraced the Gen V LT-series engine, producing swap components, such as oil pans, motor mounts, and (slowly but surely) header options. The lack of aftermarket headers has been an issue for most LT swaps. There simply are not very many options, which tends to hold swappers at bay. However, factory truck manifolds fit some applications and can be modified in others to fit. Fortunately, there are more vehicle-specific header options coming out all the time.

Outside of physically mounting the engine and transmission, there are many choices for the peripheral components. The electronic control module (ECM) for Gen V LT-series engines works quite well even in high-horsepower boosted applications, but that also requires special software that is either very simplistic or extremely complicated to use, limiting the average swapper. Aftermarket controllers are often relatively easy to tune and have the ability to get into advanced tuning parameters without special software. Basic tuning can usually be handled with a handheld programmer, which simplifies using a stock ECM, but these units do not address the vehicle anti-theft security system (VATS), which must be removed to even start the engine. ECMs can be reprogrammed by a harness supplier in most cases, and then a handheld tuner can make small adjustments.

We have spent the last three years swapping Gen V LT-series engines into just about anything, and you can find all of those procedures and much more inside this book.

WHAT IS THE GEN V LT-SERIES ENGINE?

Ever since the release of the original small-block Chevy in 1955, Chevrolet engines have been the king of all engine swaps. Some of this is due to the sheer production volume of these engines, but in modern times, aftermarket support and ease of installation has allowed Chevrolet to continue dominating the realm of engine swaps. While the LS platform remains the current king of swaps, the LT series of direct-injected V-8s and lone V-6s are primed to take over.

In 2013, General Motors released the Gen V platform, which will eventually replace the LS-series engine in all platforms. The Gen V shares the look of the III/IV series, but in reality, it is all new. The biggest advancement in the LT-series is the use of direct injection, where the fuel is sprayed directly into the combustion chamber at high pressure (2,175 psi for the LT1), which aids in fuel economy and overall performance through better fuel atomization. Direct injection also makes cylinder deactivation more

efficient, further increasing fuel economy. The 2014 LT1 Corvette can get as good as 29 miles per gallon (mpg). Other advancements include piston oiling jets, active fuel management, and continuously variable valve timing.

Because of the direct-injection method, the intake valves must be cleaned regularly—some suggest every 5,000 to 10,000 miles. This is performed with a spray-in additive while the engine is running. If this is not done, the intake valves get gunked up, causing serious drivability issues. This is the nature of direct injection.

The LT5 engine has a second set of injectors in the intake above the intake valves. This eliminates the need for the cleaning agent, but all other LT-series engines need this process. A rule of thumb is to clean them at every oil change to make sure that you don't forget. It is a required service at least every 25,000 miles, which is three oil changes because Gen V engines have a recommended oil-change interval of 7,500 miles. This is preventative maintenance

to eliminate the larger expense of a top-end rebuild.

LT Car Engines

Chevrolet Performance currently has three crate versions of the Gen V: a naturally aspirated 6.2L 460-hp LT1 (the engine installed in the base-model C7 Corvette), the supercharged 6.2L 650-hp LT4 (the engine in the Z06 version of the C7 Corvette), and the LT376, the newest LT-series crate engine that is essentially a hopped-up naturally aspirated LT1 with GM's high-lift LT1 Hot Cam and CNC-ported heads, generating 535 hp on a base tune. The LT1 for Camaros is rated at 455 hp.

Most swappers procure factory-installed engines. These powerplants have been installed in GM trucks and SUVs beginning in 2014 (1500 series only) as well as Corvettes and Camaros. The 4.3L LV3 Ecotec V-6 is the 6-cylinder variant of the LT-series, which is available in 1500-series GM trucks as well.

6.2L LT1

Making 460 hp without a supercharger is not easy, and to do so while hitting 29 mpg is even harder, but the LT1 does exactly that. The 4.06-inch bore combined with the 3.62-inch stroke creates an 11.5:1 compression ratio, which makes efficient use of the fuel pumped through the direct-injection nozzles. A forged crank, hypereutectic pistons, and forged powdered metal rods yield light weight and durability. The heads are conventional aluminum castings and feature lightweight sodium-filled valves. There two oiling systems available: a wet sump and a dry sump.

6.2L LT4

To pump up the output of the LT1, General Motors dropped a supercharger onto the 6.2L block to make 650 hp. To make that work long-term, some changes were made to the rotating assembly. The crank is the same, but the rods were slightly redesigned to increase strength in key areas. The pistons in the LT4 are forged, and the combustion chamber was opened up, decreasing the compression ratio to a boost-friendly 10.0:1. The heads are rotocast, making them stronger and better at handling higher heat ranges. The valves are solid titanium, and the oiling system is a dry-sump design, same as the LT5, and is an option on the LT1.

6.2L LT5

In late 2017, General Motors announced the release of the newest version of the Gen V LT-series engine: the LT5. This is a supercharged V-8 that is similar to the LT4, except this monster motor uses a higher-output supercharger and a redesigned crankshaft and new fuel injection system to generate 750 hp and 715 ft-lbs of torque. The most powerful GM production engine is slated for installation in the 2019 Corvette ZR1. The oiling system is dry sump only.

Gen V Truck Engines

Beginning in 2014, all GMC/Chevrolet trucks, vans, and

In 2014, the LT1 stormed onto the scene in the Corvette. This 460-hp beast features direct injection, where the fuel is sprayed directly into the combustion chamber, ensuring adequate combustion. The science behind how this works is fascinating. There was a lot of interest in swapping these engines, but the fueling system kept the swaps from taking hold at first. (Photo Courtesy General Motors)

General Motors quickly released the LT4, a supercharged version of the LT1. Basic changes were lower-compression heads (down from 11.5 to 10:1) and a big supercharger. The LT4 is good for 650 hp in stock tune. (Photo Courtesy General Motors)

Because everyone needs more horsepower, General Motors created the LT5. This beast adds 100 more ponies over the LT4, for a total 750 hp and 715 ft-lbs of torque. The last C7 Corvette, the ZR1, received this engine in 2019, and it is now available as a stand-alone crate engine. (Photo Courtesy General Motors)

While most would dismiss a V-6, the LT series has a V-6 in the lineup, and it is pretty impressive in its own right. It is 285 hp stock; a tune would easily take it to 350; add a turbo, and it could probably get into the 500s. Plus, it is small and lightweight, so it can fit in cars where a V-8 won't. This engine is found in GM trucks and vans. (Photo Courtesy General Motors)

full-size SUVs with V-8 gasoline engines came with Gen V engines. There are currently three truck versions: the LV3 4.3L (LT-based V-6), the L83 5.3L V-8, and the L86 6.2L V-8. The V-6 is an LT-series engine, essentially a V-8 with two cylinders cut off. The V-8s are the most common for trucks and SUVs.

4.3L LV3

A V-6 in a swap book? Some might balk at the idea of swapping a V-6 when they could swap in a V-8, but consider the merits of the 6-banger before writing it off. For starters, it is an LT engine with two cylinders lopped off, just like the previous 4.3L V-6, which was based on the second-gen model (also named LT1, coincidentally). The

LV3 features 11.0:1 compression with 99.6-mm bore on a 92-mm stroke, maxing the RPM at 5,800. It uses a forged steel crank with powdered metal connecting rods and caps, just like the rest of the Gen V LT family.

With 285 hp and 305 ft-lbs of torque, this diminutive powerplant has the potential to make 400 hp with minor upgrades. It could easily reach 350 hp with just a simple tune. The V-6 platform has a smaller block, which provides more options for swap projects. Fitting this 6-cylinder into say an MGB is much easier than the V-8 version, and it can make almost as much power. The fuel economy on the LV3 is 18 city, 24 highway in trucks that weigh upward of 6,000 pounds. Drop that into a 2,500-pound

Euro sports car, and the economy will be substantially better.

5.3L L83

This engine features a 3.78-inch bore, 3.62-inch stroke, and 11.0:1 compression ratio. These engines make 355 hp and 383 ft-lbs of torque with gas, while E85 produces 376 hp and 416 ft-lbs. Readily available from most salvage yards, these engines are not yet in demand because they are so new. Prices are currently under $2,000 for a complete L83, and the ECM and fuel-pump modules are inexpensive too. Once these models begin wearing out engines and the swap market picks up, the prices will go up. An LT-based engine will be cheaper than an LS Vortec, and the LTs have fewer miles.

General Motors rolled out the LT series right away in all trucks and SUVs with the L83 5.3L V-8. Having 355 hp and extremely good fuel economy make this a great engine for a swap. There are tons of them out there in low-mileage wrecked trucks. (Photo Courtesy General Motors)

You can't leave all the fun to the cars, so General Motors dropped the L86 6.2L V-8 into the high-end Denali and High Country truck models, and in 2018, the company began offering them in the high-end SUVs as well. At 420 hp with pull well into triple digits, these 6.2L L86s can do impressive things in a 6,500-pound truck. Capable of low-14-second quarter-mile times and a blistering 5.4-second 0–60 time, the L86 6.2L engines are more than capable. They are harder to find, and cost more, but they are worth every penny. (Photo Courtesy General Motors)

6.2L L86

The L86 is a modified LT1 that makes 420 hp and 460 ft-lbs of torque. The LT1 and L86 are very similar down to the compression ratio of 11.5:1. If you want any of the larger 6.2L Gen V engines, you are going to pay for it, but not as much as a 6L LS will cost. Current prices are in the $2,500 to $5,000 range for an L86 from a low-mileage wreck.

Fuel economy from the L86 is quite impressive as well. GM trucks with this engine often see 22–25 mpg on the highway, which is incredible for trucks weighing in at 6,000-plus pounds. In my personal 2015 GMC Denali 1500, a 25-mile best of 34.7 mpg was

recorded. This was under perfect conditions and in a slight downhill stretch, but it happened, and it was spectacular. The physical differences between the LT1 and the L86 are: the intake (the L86 intake is larger and makes more torque), the exhaust system, and there is the optional dry-sump oiling system for LT1s. The extra 40 hp comes from tuning and the intake. They even share the same camshaft.

Engine-Swap Projects

The goal of this book is to assist in an LT-series engine swap. Whether swapping a 1969 Chevy truck, a 1970 Chevelle, or

a 1999 Miata, the information in this book will help you achieve your goals. Performance is typically the number one goal of any engine swap, and the LT-series offers that in spades. As these engines become more popular, the aftermarket is rapidly producing more performance components. Spicing up an LT engine is almost as easy as ordering the parts themselves.

The main concerns for LT swaps are fitting the oil pan, accessory drive, power steering, and fuel system. Another issue is the exhaust (mainly headers or manifolds) because the head design for LT engines is different from the LS series, and there are simply not very many options for swaps, so you need to get a little creative. Luckily, there are more options now than there were six months ago, so by the time you read this book, there will be even more options available for your LT swap.

Locating an LT engine is as simple as ordering a crate engine from a dealer, parts house, or local salvage yard. There are distinct differences between the car and truck engines. All truck engines have a longer crank pulley, which is because the truck engines use an engine-mounted belt-driven vacuum pump for the brake assist. Additionally, the truck engines use a driver-side biased water pump (all LT pumps are offset), while the car engines use a passenger-side offset pulley. This is not a big deal for most applications, but the accessory drives can't be interchanged without swapping all of it.

Swapping Problems

The LT-platform has tons of potential for increased performance, but there are some caveats that must be addressed for swaps. The main issues are the fuel system and the lack of a power steering pump. Both of these are addressed at length in the pages of this book, but it is something that you need to know going into planning an LT swap.

Nearly every new vehicle uses electric power steering. This reduces drag on the engine and gives the manufacturer the ability to tune the steering assist based on vehicle speed. There are options for aftermarket electric power steering, such as with American Powertrain, otherwise an aftermarket accessory drive can be used on an LT engine to have traditional power steering. This is the main reason that General Motors is not using the LT engine in the larger 2500- and 3500-series trucks. These vehicles use a hydroboost for the braking system, and they simply need to have a power steering pump to provide the hydraulic pressure for the hydroboost.

The fuel system is the other major departure from the traditional swap. LT engines are direct injected, which uses a PWM fuel pump without a return line to feed the engine with up to 76 psi of fuel pressure. From there, the engine further increases the fuel pressure with a mechanical fuel pump. The chassis or tank pump is therefore a lift pump, essentially moving the fuel from the tank to the engine. The complexities of the fuel system are addressed in chapter 8.

Outside of these two areas, an LT swap is fairly simple. There are some notable differences from other engine platforms, but that is to be expected. The rest of this book deals with how to perform a swap and covers most of the details. While every car and swap are different, there are quite a few common aspects.

Swapping an LT engine into just about anything is not the most complicated automotive endeavor. An average swapping project is fairly easy if it is carefully conceived, researched, and planned. The Gen V LT engine shares a similar footprint with the original small-block Chevy. The general rule of thumb is that if a small-block fits, an LT will fit as well, with some minor adjustments of course. Engine position, oil pan, and accessory drives are the most common physical fitment issues.

Unlike previous fuel-injected engines, there is no carbureted option for LT-series engines. The nature of direct injection prevents the possibility of using a carburetor. This means that every LT swap requires using an electronic control module (ECM) and sensors. In most cases, you must carefully modify the wiring harness, plugs, and wiring, or purchase the correct aftermarket components for plugging in the particular engine to a specific car. Chevrolet Performance, Howell EFI, HP Tuners, and many others offer products and tuning that make swapping the electronics much easier. In the end, you get a more efficient powerplant with the ability to tune it better and faster.

One of the most affordable ways of procuring a GM LT-series engine is through a salvage yard. Because these engines are so new, the demand is quite low, but people wreck trucks every single day. When a new truck is totaled, it goes to a salvage yard. While body panels and interior pieces are in demand for repairs, the drivetrains are so new that there just are not very many on the road with enough miles on them to break down. In fact, many of these vehicles are still under warranty. This means the market is in the prime position to buy low-mileage LT-series engines at a substantial savings. Within the next five years, the price of these engines will go up considerably.

If you have never purchased a salvage engine, you may be a little leery of the process. While you can certainly be taken advantage of, the more you research, the better off you will be. There are several keys to successfully buying a salvage engine: knowing where to buy, knowing what you are buying, and finding what parts are available.

Where to Buy

Knowing who you are buying from is just as important as knowing what you want. The internet is a glorious tool for helping weed out the undesirable salvage yards that take advantage of their customers. Salvage yards often have a reputation for being sleazy, and those certainly still exist, but the more-reputable yards get better reviews online, helping you make a more informed decision on where to buy. There are some salvage yard networks, such as LKQ, that link quality yards together so that you can find the parts you need from all over the country.

Then there is the pull-a-part-style yard, where you remove the parts that you need. These yards are usually less expensive.

Inside the crate is a brand-new $8,500 to $10,000 engine, all shiny and clean. Don't want to spend that much? There are other options. Want to spend more? The LT5 is now available as a crate engine, and you can also order a complete drivetrain with the ECM, transmission, and all the controllers, harnesses, and other components needed for the drivetrain itself. This does not include the fuel system, mounts, or accessories.

When ordering a crate engine, this is what you get: a big box on a pallet. You will need a forklift or pallet jack. This LT1 crate engine was ordered with the complete ECM control package and was installed in the 1971 Buick GS seen in this book.

However, you need to research the parts, find the vehicle, and remove the parts using your own tools. Some yards provide cherry pickers or forklifts to remove large parts, such as engines, while others do not. That is the caveat emptor of salvage yards: research the yard and ask questions.

Know What You Are Purchasing

Purchasing a used engine means dealing with a shop or salvage yard that has dismantled a vehicle. When it comes to engines, most yards want to get the job done as quickly as possible and break the vehicle down to as many sellable parts as possible. This means that for most engines, the wiring harness gets chopped along with the fluid lines. Most yards remove the peripheral components, including the throttle body, alternator, starter, and air-conditioner compressor. Some yards strip engines to the long-block, meaning no intake, exhaust manifolds, or water pump as well. It pays to ask what comes with the engine when you are researching your purchase. You might pay an extra hundred or two for a complete engine, but buying the peripheral components can cost thousands in the end.

If at all possible, get the ECM and throttle pedal from the same vehicle. It is not absolutely critical, but it is nice to know everything is already paired together.

Get the Vehicle Information

Because there are some key differences in the engines themselves, you need to know where your engine came from. While the original vehicle's vehicle identification number (VIN) is not as important, get the year, make, and model of the vehicle, along with the original mileage. There are small differences between year, make, and model for each engine, and you need that information readily available for future service.

Salvage yards leave the harness plugs attached. Most of the time you can't get a salvage harness, because it takes too long to strip it out, so they just chop the wires. We have tried requesting a used harness, but they always tell us they won't do it. This varies yard to yard. All the sensors are there.

Parts That Are Available

Buying a used engine, whether it is from a salvage yard or a private seller, typically means you are getting part of the package but not the complete package. With LT engines, you need the engine, throttle body, throttle pedal, ECM, wire harness, fuel control module, and accessories. If you can source this all from the same vehicle, that is perfect, but if not, you need to match the components from a similar vehicle and engine.

A V-6 throttle body is different from the 5.3L L83, and the 6.2L L86/LT1/4/5 engines use a different throttle body than the smaller Gen V engines. You can convert to the larger throttle body, but swapping a 5.3L throttle body to a 6.2L will decrease performance. The ECM, fuel modules, and wiring harnesses are the same for the truck engines.

As with any purchase, the more you can research on the seller and the parts you are buying, the better off you will be. Keep in mind that any used engine is just that—used. There are no warranties from General Motors, but many salvage yards offer short-term warranties and even extended warranties that can be a great benefit, which is just one more aspect to consider when shopping for a used LT engine. ∎

This is what $1,500 will get from just about any larger salvage yard: an engine without any accessories or wires. We scored this one from LKQ, a national chain of salvage yards, for $1,575 with free shipping because it was local. You don't get the throttle body, and usually you don't get the exhaust manifolds either. We were lucky. This take-out engine found its new home in a 1987 Camaro built for this book.

ENGINE AND TRANSMISSION MOUNTS

The first thing that needs to be done in a swap is physically fitting the engine into the chassis. This may require several attempts, so make sure to have a quality engine hoist, heavy chains (or better yet an adjustable load leveler), and a helper. It is much easier and safer to do this work with help. Make sure to have a level, measuring tape, and notepad for taking measurements and notes on the fitment into the chassis. If the original engine is still in the chassis, takes photos and measurements for reference.

Installing an LT engine begins with engine mounts because this is the key to everything else. There are many ways to get this done, depending on the vehicle. A bolt-on adapter is the most common and by far the simplest solution.

Motor Mounts

Much like the LS platform, the LT-series engines use a four-bolt engine mount that is just forward of the block centerline where the rear two bolts split the center cylinders. A small-block Chevy (SBC) has mounting bosses on the block that sit roughly between the two front cylinders. This pushes the LT forward of where the typical SBC sits. Adapting the LT to fit where an SBC or LS rested is not much of an issue; retrofitting an LT to a platform other than that is a little more challenging, but even that is pretty simple.

Dimensionally, the LT is about the same size as the SBC, although the back of the block is shorter than the SBC because there is no ledge for the bellhousing. Additionally, the mechanical fuel pump is at the back of the block, requiring extra clearance. Most swappers simply adapt the LT-block mounts to the commonly available SBC mounts and then go from there. With so many options for placement, it can get a bit sticky trying to figure out exactly what is needed.

The factory motor mounts are a wild departure from the standard SBC style. There is a large damper on a large three-bolt flange. Shown here is a 2014 L83 engine with truck motor mounts.

The factory LT engine mounts are wildly different from SBC-style mounts. These mounts use a large polymer bushing to dampen the engine vibrations. The bushing is bolted to the frame, as opposed to a clamshell that is secured with a single through-bolt. In some instances, the factory mounts can be used, but they are very large and require substantial fabrication to get the alignment right. Using SBC mount adapters offers more options, and they are generally easier to fabricate frame stands for.

Typical installations need the engine in the factory SBC location, which references the bell-housing mounting plane. This is 1⅛ inches rearward from the factory LT mounting point, so these are known as 1⅛-inch rearward adapters. This position works well in most GM vehicles but not all. The 1987 Camaro project featured in this book required setting the engine, an L83 5.3L truck LT-series engine, just 5/8 inch rearward to gain the necessary clearance for the firewall. We used adapters from Trans-Dapt, which also makes the 1⅛-inch rearward

adapters and 5/8-inch forward pieces for vehicles that need even more clearance.

Adapting an LT to a factory GM SBC frame stand is a balancing act. There is the firewall clearance to deal with as well as the front of the engine. The oil pan and accessories can get in the way, especially when using factory components. In most cases, the factory truck-only vacuum pump will not clear any GM muscle car frame, but it will clear on most truck chassis. Additionally, the factory AC compressor location is on the passenger's side, down low, tight to the block. Many aftermarket adapter mounts simply will not clear the factory compressor, and the compressor does not clear the frame or crossmember in most vehicles, so you have to sort that out. We have a solution though in chapter 4.

Adapter Plates

Most swappers elect to use adapter plates because they are simple and allow the use of readily available (and inexpensive) SBC mounts. There are

many options for adapters from single-position adapters, which are the most common and simple, to multi-position and even sliding mounts. We have used all three types, and they have their benefits and drawbacks. The biggest issue with single-position adapters is that they are just that: one position. If you run into a clearance issue, purchase a different set that works with your project.

There are substantially fewer companies making Gen V adapters compared to LS adapters, which means fewer are available to choose from. This isn't really a big problem, however, because those that are available are quite versatile. Brands including Trans-Dapt, ICT Billet, and Dirty Dingo have the parts needed. For those swapping LTs in the place of LS-series engines, ICT Billet even makes Gen V LT-to-LS chassis adapters.

Trans-Dapt

If basic adapters are needed, these are the ones you want. They are built from laser-cut steel and painted black, and they work. They come in three flavors: 1⅛ inch rearward, 5/8 inch rearward, and 5/8 inch forward (with or without rubber or polyurethane engine mounts). We used these in the 1987 Camaro featured here. They fit well, clear all of the components, and do precisely what they are supposed to, all for less than $80 (plates and hardware only, no mounts).

The measurements reference the relative position of the engine to the factory LT position, not the mounts themselves. So, 5/8 inch rearward means the engine itself sits 5/8 inch from the LT-series mount, which is half the difference between the SBC and LT positions.

Let's say you have a custom chassis or something that is not already fitted with GM SBC frame stands. In this case, custom mounts are needed. Trans-Dapt has you covered with the 4604 and 4605 biscuit-style mounts. Similar to the early SBC front mounts, these adapters are universal for frame rails that are 27 to 33 inches (4604) and 24 to 30 inches (4605) apart, so the tabs can be bolted or welded to the chassis, and then the LT can be simply dropped in place. This greatly simplifies the process of swapping an LT into non-GM vehicles. Just about any non-GM application can make use of these mounts for around $115.

This block adapter can use any of the three main GM motor mounts types: tall/narrow, short/wide, or the 1973–1998 clamshell style. This mount sets the engine rearward 5/8 inch, which is half the distance between an SBC bellhousing plane location and the LT plane location. The Camaro featured in this book used this measurement because the SBC position is too close to the firewall.

If you need to push the engine forward, this might be the mount for you. Moving the engine 5/8 inch forward from the LT bellhousing plane is a full 1¾ inches forward from the stock SBC bellhousing plane location.

If you have the room, these mounts position the LT engine bellhousing in the same location as a stock SBC engine. This is the preferred option if you have a deep firewall. These are great for trucks and most GM muscle cars.

Solid adapter plates, like this Trans-Dapt unit, clear the factory components, even though you should always delete the factory vacuum pump (these have a high failure rate and are not necessary).

The mounts use a hex-head bolt for the rear mounting points and flush socket-head bolts on the front to clear the motor mounts.

You can order the adapters from Trans-Dapt with Prothane motor mounts, which simplifies the process. They bolt on just like any other mount.

We set the L83 truck LT engine into the chassis of a 1967 Camaro to test the fit. You can see that the intake will not clear the factory hood immediately.

The only thing keeping the engine from dropping into place is the factory vacuum pump, which should be removed anyway.

ICT Billet

Say that you need an LT because you already swapped an LS into your car but you want the latest and greatest in powertrain technology, or maybe your factory LS-powered car is just not up to the task anymore. Well, my friends, the geniuses at ICT Billet have the part you need: billet aluminum LS-to-LT swap adapters. These adapters relocate the LT engine to the LS position in fine style. They clear all the important components and come with the hardware to mount them.

ICT Billet also has multiposition mounts for LT swaps. This means there are three unique position options with a single adapter. All positions move the engine 0.72 inch up in the chassis, which may not work for those using the very tall truck

These billet aluminum adapters from ICT Billet are drilled for multiple positions from the stock SBC position to 2.25 inches forward.

The frame stands determine which style of motor mount is needed. This one is on the 1971 Buick GS, and it is not compatible with any of these adapters because it is a Buick frame stand. Oldsmobile and Pontiac each have their own versions, so you need to convert to Chevy stands.

intake, but this provides clearance for the crankshaft-to-front crossmember clearance, which is a problem in some vehicles. The three positions are set at the stock SBC position (1.125 inches back), 1.625 inches forward, and 2.25 inches forward. This measurement refers to the actual engine position.

ICT Billet states that its adapters will allow the use of the factory AC compressor and vacuum pump in the 1.625-inch position with the plates trimmed to remove the factory SBC position mounting holes. Although, this may cause a clearance issue between the oil pan and steering linkage or front crossmember.

Dirty Dingo

Another option is to purchase mounts from Dirty Dingo, which has two styles: static multiposition and sliding mounts. We used the sliding mounts on a 1971 Buick GS, and they worked very well. They allow the position of the engine to be adjusted to fit the original Buick 350 frame stands, which is something that no other mount will do.

Multiposition

Like the other multiposition mounts, these start with the factory SBC position, which refers to the bellhousing mounting plane. This is the most rearward position for the engine using these mounts. From there, the mounts move 11/16 inch forward, $1^{11}\!/_{16}$ inch forward, and 2¼-inch forward from the stock SBC position. These mounts do not clear the factory AC or vacuum pump. They use multiple positions drilled and tapped into a single mount, so the position of the engine can be adjusted as needed without buying new mounts.

For unique circumstances, the sliding mounts may be the best bet. These mounts allow the engine to be positioned in the stock SBC location, up to 2 inches forward, and anywhere in between. This means that minor adjustments can be made for specific applications, while static mounts can't be adjusted. These

On the 1971 Buick GS convertible featured in this book, we used these sliding motor mounts from Dirty Dingo. They were the first commercially available LT swap mounts, and they provide some options that you don't get with other adapters. Instead of a single position or several fixed points, the adapter bolts the block, while the motor mount bolts to a second piece that slides on the block adapter. This allows you to put the engine wherever you need it. These are very large though, so they do interfere with the factory AC compressor mount.

An impact gun and a backup wrench were used to remove the old mounts from the frame.

Luckily, all GM frames are drilled for multiple engine options, so the SBC frame stands just bolt right on, using the correct placement.

The Dirty Dingo mounts are machined to accept the SBC mount, which has a bubble on the back. Not all adapter mounts have this, which requires grinding the bubble away.

This is the bubble, which is more commonly found on aftermarket motor mounts. The spacer plate is required.

Once assembled, the adapter slide is ready to go on the engine block. Note the recessed fasteners, which make servicing the engine mounts more difficult.

The slider mounts to the engine block adapter, allowing the engine to be positioned from the stock SBC plane to 2 inches forward. Unfortunately, these mounts are really large, so they don't clear the oil ports on the driver's side.

mounts are quite long and a little complicated to secure, but they are worth it if you are working with nonstandard frame stands, such as Buick, Olds, and Pontiac. They do not clear the vacuum pump or AC compressor using the factory accessory drive.

GM Truck Mounts

Standard adapter plates work for GM two-wheel-drive (2WD) trucks with SBC frame stands, but four-wheel-drive (4WD) trucks are a much different story. The crossmember is in the way, which keeps the standard mounts from working. Dirty Dingo offers 1973–1999 Chevy truck 4x4 mounts that are adjustable from the standard SBC position up to 2.5 inches forward and an additional 1/2 inch rearward. These are made from laser-cut steel and are powder coated black. They do not clear low-mount AC but will clear the vacuum pump.

SBC

Most adapter mounts are designed to work with the standard SBC three-bolt engine mount. For GM-chassis vehicles, there are three different versions of motor mount: tall/narrow (early style), short/wide, and clamshell. The first generation of SBC mount was a biscuit mount used on 1955–1957 GM chassis. In most cases, these chassis need to be converted to a later-style engine stand.

Clamshell

The clamshell type is more common on later GM vehicles, but they can be found on some

These are the common SBC-style mounts that most LT-swap adapters use. There are two versions of these mounts. The most common is the tall/narrow (left), which are used in most GM applications. On the right is the short/wide mount, which are used on GM 307 SBC applications. Determining which one is needed depends on the available frame stands.

The center of the tall/narrow engine mounts measures 2⅜ inches between the mounting ears and 2³⁄₂₆ inches tall (from the center of the mounting bolt to the top of the engine mount pad). The short version (right) is 1¾ inches tall and 2⅝ inches wide.

converted 1964–1972s. The clamshell's design uses a stamped-steel pod that bolts to the engine with a steel and rubber mount that bolts to the frame. These can be used with most adapter mounts.

Tall/Narrow

These are most common in Chevrolet muscle cars. However, the tall/narrow distinction is confusing because the frame pad is called short/narrow. The tall/narrow refers to the engine-mounted component. These are the SBC mounts used for 350/396/454 engines in Chevrolet vehicles. The center of the engine mounts measure 2⅜ inches between the mounting ears and 2³⁄₂₆ inches tall (from the center of the mounting bolt to the top of the engine mount pad).

On the frame stand, the mounting pad measurements are 2⅜ inches wide and 1⅝ inches tall (crossmember to pad). GM part numbers for these frame stands are 3980711 for left hand and 3980712 for the right hand. These frame stands are readily available in the aftermarket as reproductions.

Short/Wide

For the Chevy 307-ci engine, General Motors used a different set of frame stands. The frame stands are taller than the 350 version by 1/2 inch. The width of the pad (where the two mounts come together) is also different; the 307 mounts are wider, measuring 2⅝ inches on both the frame stand and the engine block mount. The block mount is shorter than

The limits of the sliding mounts with the factory accessory drive are shown here. You can move the engine forward but not fully rearward to the SBC position.

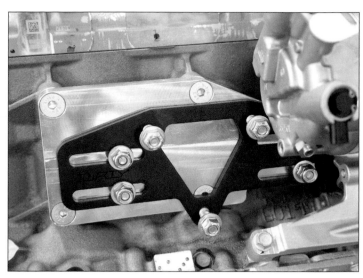

These mounts are really nice, but they do have some flaws. They are complicated to assemble and service. Adjusting the position of the engine takes a lot of effort because the fasteners quickly become hard to reach.

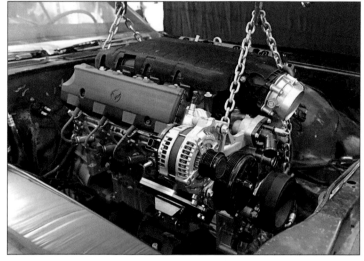

When setting an engine into the engine bay, use a load leveler like the one shown here. This allows you to shift the engine as needed without repositioning chains. The tighter the engine is to the firewall can be an issue though because as there must be space to remove the bolts in the rear of the engine.

One of the biggest issues with LT1 swaps is the water-to-oil cooler. In most applications, this simply does not clear the frame. Most GM trucks have enough clearance, but car frames do not.

the 350 version, measuring 1¾ inches tall. These are the most commonly sold mounts at the parts store, so it pays to know the difference.

The type of mount needed depends on the vehicle, the accessory drive, and the oil pan.

That being said, the best solution is to use the 350 version. These mounts raise the engine a little higher than the 307, which provides better clearance for the oil pan and steering linkage. In most cases, the engine will still need to be raised a little more to clear

the steering linkage. About 1/2 inch usually works, depending on the oil pan and the angle of the engine/transmission.

If working with a non-SBC-powered GM chassis (such as a Buick, Olds, Pontiac, or Cadillac), then convert to the

Certain objects that can't be moved might be in the way, so you have to be able to adjust. With the 5/8-inch rear adapters, the engine clears the wiper motor on the 1987 Camaro when installed, but it does not clear during the install process, so you have to remove it.

To install the engine into the chassis, we had to drop the rear of the engine, lower it, and then drop the front of the engine, and then lower it some more. The leveler gets a real workout in the process.

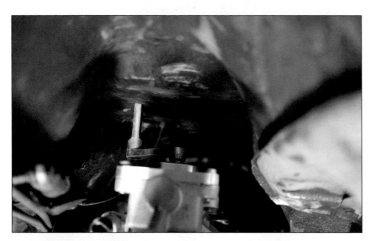

Once the engine is in the car, start on the transmission. The Camaro gets a Tremec T56 Magnum 6-speed, so cut the floor to clear the shifter before mounting the transmission crossmember.

Remove the factory automatic shifter, which requires unbolting the shifter.

Next, drill out some spot welds to remove the bracket. This bracket must be removed before cutting the floor.

We used a small Sawzall to cut the floor. Several cuts may be needed to get the hole to the correct size and shape. Cut minimal material to reduce drafts.

This was the first section we made. This hole eventually became a few inches longer to notch the floor for the shape of the shifter pad on the 6-speed.

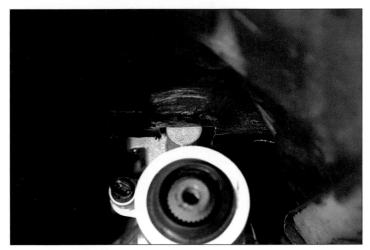

Here is the notch for the rear of the shifter pad.

The stock crossmember will not work at all for the T56, and there was not an LT-specific transmission crossmember for the third-gen F-Body at press time, so we picked up a Holley T56 crossmember and modified it. The LT sits in a different place than the SBC or LS engine would, so the crossmember does not line up. You can see the T56 mounting pad hits the crossmember itself.

We notched the pad and the crossmember to clear, removing only what was necessary.

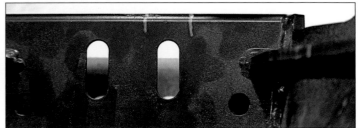

The body of the T56 hits the base of the crossmember as well. We marked the interference so we could trim this away.

Now the crossmember fits. We could probably leave it like this, but we wanted to gain the strength back. So, we cut a gusset to box the base together again.

Now the T56 bolts down in the correct position and is secure.

We considered modifying the transmission crossmember, and actually attempted it, but the results just didn't work, so we opted to buy a new one.

Because the transmission sits so far back, the original slots for the trans mount didn't align. They were enlarged—both top and bottom. You can see where the section that was cut away was closed up.

For LT swaps with automatics, you will have to fight the transmission crossmember as well. In the 1971 Buick GS, we used a 4L65 automatic, which bolts to the original crossmember. However, the position of the engine did not align with any of the factory mounts for the crossmember. The crossmember sits too high in the car to allow for adequate clearance, and the transmission angle is way too high (we need the angle to be closer to 2 degrees down).

The GM A-Body crossmember from G Force fit well and worked, but the trans mount tab didn't clear for our positioning with the LT1, so G Force redesigned the crossmember.

Instead of a tab, the newly designed unit uses a shelf, which works much better for the LT position in the A-Body. This crossmember is very heavy, but it is solid and fits quite well.

To install the G Force unit, we had to drill out the original bolt holes in the frame. The new crossmember uses larger bolts than the factory unit.

Getting the nuts on the bolts is a bit of a challenge because the holes in the bottom of the frame are too small. We pressed the nut (nylock style) onto another bolt and used that to help thread it onto the bolt. This is an A-Body convertible/big-block issue only; standard SBC coupes have a different frame. All Buick, Chevy, Olds, and Pontiac A-Body convertibles and big-block models use the frame shown here.

Because we used sliding mounts on the engine, we need a sliding mount on the transmission as well. This piece came from G Force to match the crossmember, and it is the only one of its kind. It works very well, and it can be purchased without a crossmember. If using sliding mounts, use this trans mount as well.

We needed some adapters to get the driveline angle correct. They are included in the kit.

It is complicated to install the mount because it must be done in a particular order. First, the mount bolts to the transmission, and then the lower half installs between the trans mount and the crossmember.

OIL PANS

Regardless of the vehicle, the oil pan is often the most difficult obstacle to overcome in an engine swap. Unlike the LS platform, which has no less than 10 oil pan variations from the factory and countless aftermarket options, the Gen V LT-series engine only has four factory options, none of which are well-suited for swaps, and the aftermarket is just now offering proper solutions. As with anything, a little research and fabrication helps move things along.

From the start, you need to know that modifying an LT oil pan is a nonstarter. Unlike LS pans, the LT's oil pickup tube is built into the pan itself; it is cast into it. This makes chopping out sections of the pan to clear obstacles not possible like it is on the LS

What is certainly the most popular LT-swap engine, the L83 5.3L truck engine has a unique oil pan design, which uses a cast pan with a metal bolt-on sump. Additionally, some L83 and L86 truck engines have an air-to-oil cooler system, as shown here by the aluminum block and lines coming from the pan. It can be used or deleted.

pans. Fortunately, the aftermarket has solutions. Sorting out what works and what doesn't depends on many variables, including the chassis setup, engine and transmission mounts, intended use for the vehicle, and the engine itself.

Factory Oil Pans

There are four factory pans: truck, LT1 wet-sump car, CTS-V/LT4 wet sump, and LT1/LT4 dry sump. Eventually there will be others, but these are the current GM factory options. They are all cast aluminum with integral pickup

tubes and oval oil ports to the block. This means that you can't simply cut and weld the oil pan. The factory oil pans all have a fairly large sump, which is far forward on the pan itself. This means that in most full-frame car applications, the factory pan simply will not fit. Most 2WD trucks will accept the factory car wet-sump oil pans without additional modification, but the truck pan does not fit.

Truck Pan

The truck pan is quite large. Not only is the shallow section of the pan very short, there is a sheet

metal extension pan that bolts to the bottom of the main sump. It is very deep, which makes it impractical for use in car swaps. The truck pan will fit 1972-and-older trucks without modification, but 1973-and-newer trucks will require special motor mounts and heavy crossmember modifications. Oil capacity is 8 quarts.

LT1 Car Wet Sump

Found on most LT1-powered cars, the standard wet-sump oil pan is shallow, measuring just 4.75 inches from the bottom of the block. The main sump is 15 inches long and measures 4.75 inches at the back to 4.5 inches deep at the front. The shallow section is 7 inches long and tapers from 2.5

inches deep at the rearmost section to 1.5 inches at the front of the pan (front of engine). This pan will fit most trucks, but most cars will require crossmember modifications because the main sump section is too far forward to clear. Oil capacity is 7 quarts.

2016–2017 CTS-V/LT4 Wet Sump

This pan is very deep, making it unlikely to be suitable for most

The high-powered CTS-V/LT4 wet-sump oil pan is quite large. It may work in some trucks, but the pan is very deep, 7 inches at the deepest section of the sump, so it is not a good option for lowered vehicles.

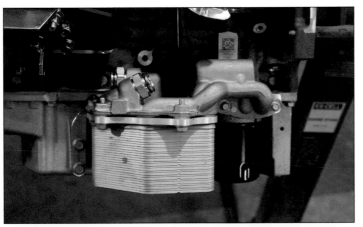

This is the LT1 wet-sump oil pan with the water-to-oil cooler system installed. The oil cooler does not fit most chassis designs.

On the passenger's side of the LT1 wet-sump pan is an oil level sensor. The overall shape of the pan is long, preventing this pan from working in most frames without modification.

swaps. The shallow front section is 11 inches long, so it may clear some stock crossmembers, depending on the engine setback. The shallow section is 1.5 inches deep at the front, tapering to 2.5 inches at the shelf for the main sump. The 7-inch-deep sump is 12 inches long. The main issue with this pan is that it is quite deep, leaving it exposed to road debris and potential destruction. If the vehicle is lowered at all, this pan should not be considered. Oil capacity is 10 quarts.

LT1/4 Dry Sump

Certain vehicles and crate engines come with an optional dry-sump system. The pan for this is similar to the wet-sump car pan, but the shallow section is a little bit shorter than the wet-sump pan, making this a difficult swap for most full-frame vehicles without modifications to the crossmember. This pan has two oil drain plugs. Oil capacity is 9.8 quarts.

Modifying the crossmember to clear a factory oil pan is not complicated, and while it does require good quality fabrication, it is a fairly simple process. The key to this is taking plenty

Unlike the other pans, the CTS-V/LT4 wet-sump pan does not have oil pressure port options, making it even more difficult to use for swaps.

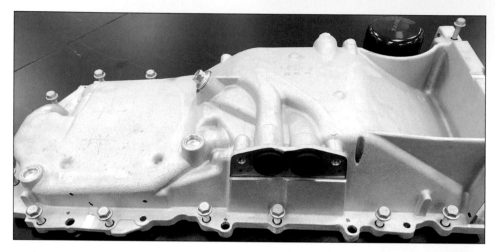

The LT1/LT4 dry-sump system is very nice, but if it is chosen for your swap, chances are good that the cross-member will need to be chopped up to make it fit. This is very similar to the wet-sump pan, only with a shorter rear section. The front is just as deep, which will interfere with most frames.

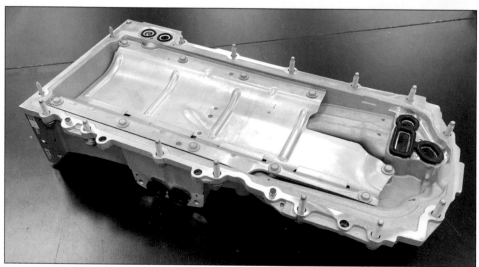

The inside of the dry-sump pan shows how difficult it would be to modify the pan itself. The pickup lines are cast into the pan, so essentially the whole thing has to be redesigned.

of measurements with several reference points that will not change. This allows you to place the engine without the oil pan installed, make measurements for clearance, and extrapolate the necessary notch required for your application.

Aftermarket Oil Pans

It took the aftermarket a few years to get going on LT-swap components, but it has finally come around and there are several LT-swap pans available. Currently, there are four oil pans available for the Gen V LT-series

engine: the Holley Retro-fit (standard and drag race versions), the Moroso fabricated pan, and two pans from BRP HotRods—the High Clearance and Extended Sump. All of these pans are designed to add clearance for LT-swaps and should work for most applications. The details set them apart—as does the price. All of these pans are designed for wet-sump oiling systems.

Holley Retro-Fit Street and Drag

Holley has been at the forefront of the LS swap game for quite some time, and things are no different when it comes to the LT-series engines. Like its LS pans, the LT Retro-fit Street pan (part number 302-20) is cast aluminum, has a built-in oil filter port that matches the factory placement, and has all the other fittings seen on a factory pan.

The pan reduces the rear sump length by 5 inches, providing

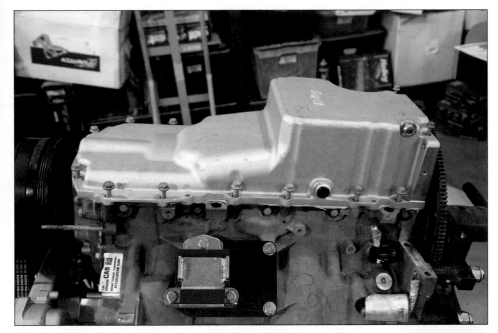

Aftermarket oil pans are now available, which is a really good thing. This Holley swap pan fits like a dream and has all the same features as the factory pans, but this one actually fits most chassis designs. The Holley pan is the most affordable of all the currently available options at around $400.

This is the BRP HotRods LT swap oil pan, which is a fabricated pan (welded sheet aluminum instead of cast), which is quite nice. It fits the block, has all the right ports, and is designed to work with the BRP LT swap system.

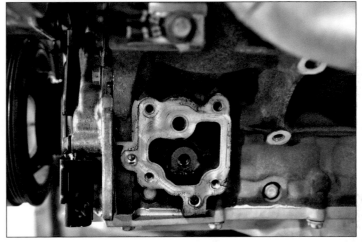

All of the truck LT engines up to 2018 use a block-mounted vacuum pump to provide extra vacuum for the braking system. These do not clear many frames and are subject to a massive recall by General Motors because they have been failing and causing braking issues. The upper port is pressurized and the lower port is a drain. These must be plugged or used with a fitting. The threads are 12 mm x 1.75. GM part number 11546665 is required as the plug.

The water-to-oil cooler system seems like a great idea, unfortunately it just doesn't clear most chassis designs. The cooler has two lines for water flow: on the left top of the cooler, the line facing rearward goes to the large port on the block (just above the QR code on the rear of the block), and the front-facing port runs to the radiator. Deleting this cooler requires a block-off kit.

Modifying a Crossmember for Oil Pan Clearance

Unless you purchase an aftermarket oil pan for the LT-series engines, installing one into most full-frame vehicles requires modifying the front crossmember. Each install requires different dimensional cuts, essentially cutting the crossmember to match the particular installation parameters, including the height and lateral location of the engine itself.

To begin, the engine must be placed in the vehicle. The best method is to use an engine hoist with an adjustable-angle attachment mounted to the motor. Lower the engine as far as possible until it hits. In the case of the 1971 Buick GS swap, a set of Dirty Dingo sliding motor mounts was used. This allowed the engine to be set on the mounts in the rearmost position. Having the motor in the car allows you to mark the crossmember for the oil pan location, removing much of the guesswork.

Use a permanent marker to mark the crossmember, and make a few measurements. The sliding mounts adjust 2 inches, so the front of the pan could move as much as 2 inches from its current location. To accommodate for engine

1 *With the engine in position, you can see where the pan hits the chassis. We used a gold Sharpie to outline the pan on the crossmember and then removed the engine.*

3 *Use whatever method you want to cut the metal. We used a plasma torch to make quick work of it.*

4 *Inside the crossmember, we found some scale and rust. There was nothing that would degrade the integrity of the frame, but we took the opportunity to clean it up, kill the rust, and paint it to protect it from future rusting.*

2 *The Sharpie marks the outline of the pan. Don't cut right on the lines, add at least 1/4 inch to get suitable clearance. We added 1/2 inch of clearance to have plenty of room for adjustment.*

5 *Once the jagged edges from the plasma cuts were cleaned up, we moved on to the process of making the boxing plates.*

6 *We used cardboard to trace the shapes of each section for the boxing plates. We set the depth to the lowest section of the crossmember, specifically the rearmost edge. Each pattern was then copied to 1/8-inch plate steel.*

7 *The plates were cut out using a band saw, but any other cutting method can be used. Don't forget to mark each piece because they are side specific.*

twist and movement, we added a minimum of 1/2 inch clearance on all sides of the pan. This provides ample room for adjustment when installing the engine as well.

There are several ways to cut out the steel. We chose to use a plasma torch for clean cuts, speed, and ease of use. An angle grinder could be used with a cutoff wheel as well. Once the offending steel was removed, the engine was reinstalled to check the fit. This time, the motor set in place with the mounts slid fully forward, and the oil pan had plenty of clearance.

The crossmember has now lost a good deal of structural integrity because a large section has been removed. If this was left as is, the crossmember could flex or even collapse. To put the strength back into the crossmember, 1/8-inch plate steel was used to box in the notched section, eliminating the potential flex points.

The best method for this process is to use poster board to create patterns for each piece. Decide whether or not to place each piece on the inside or outside of the cut. Going on the outside will reduce the clearance and requires a better initial cut of the crossmember. Going on the inside yields a better fit and finish, but it also requires a more precise cut on the new pieces. We chose to go to the inside for a better finish.

Another option is to leave the floor of the crossmember open. We did not go this route, preferring the fully boxed application. We made a new flat floor to complete the box work on the notch, which more closely resembles the factory look and design.

Once the templates were made, they were traced onto the steel (1/8 inch) and cut out using a band saw. A plasma torch or a cutoff wheel can also be used. The band saw allows for a more accurate cut. Each piece gets labeled for placement as it is cut so that it will not get mixed up with any of the other pieces.

We used a grinder to clean up the edges of the crossmember before the final fitment of the boxing. Plasma cutting leaves slag on the edges. Clearing off any jagged edges makes for a better fit and less cleanup after the welding is done.

The welding step takes patience. The pieces may not fit together perfectly; trim or adjust the placement slightly as you go. Use small welding magnets to hold each piece in place, which can help you put all the pieces in at once to check the overall fit. Once you are satisfied with the results, begin welding.

We tacked each piece in place with one or two tack welds. Small gaps (less than 1/8 inch) are okay here and

Modifying a Crossmember for Oil Pan Clearance *continued*

8 There may be a little trimming and adjusting to get each piece to fit perfectly. Once set, each piece was tacked into place.

9 The side plates are trickiest because several angles need to be matched up. Just a light tack weld is needed because fitment may need to be tweaked.

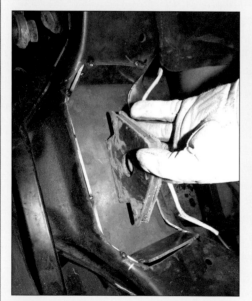

10 We used a welder's magnet to hold the bottom of the boxed section in place and tacked it in as well.

11 After a few stitch welds to hold all of the pieces together, we used a hammer to convince each side piece to conform to the natural shape of the crossmember.

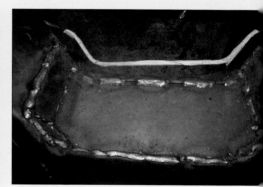

12 Then we finish welded the entire box with 1-inch-long stitch welds to reduce warping. Continue the welds until all of the seams are fully welded.

there, but big gaps or separation between the pieces or the crossmember itself are not. The floor of the box was laid into the opening before all of the side pieces. After each piece was tacked in place, the floor section was lifted into position with a magnet and tacked as well. With all the tack work done, the boxing was fully welded with a MIG welder.

The steel is thick enough to run longer beads, so we welded each section in 1- to 2-inch beads. There were a couple of problem areas where the flat plate didn't match the contour of the original crossmember. This was remedied with a few hits from a ball-peen hammer. This is necessary on most fabrication jobs like this. To keep the warping down, the welding was moved to opposite sections for every couple of inches of weld.

After the welds cooled, a flap-style grinding pad was used to dress the welds. Be careful not to dig into the metal or grind too far into the weld, weakening the seam. Dressing the welds not only makes the modification look better but also helps you find any pinholes or sections that were missed while welding. To fix these issues, just hit it with the welder and dress it down.

Once you are satisfied with the way the welds look, move on to installing the engine. On our Buick swap, the welds were dressed flush and smooth, but there were a few spots that need attention. This car is slated for a frame-off build where the details will be addressed. ■

14 *The finished boxing is shown here. This car is getting a frame-off build, so once all the mock-up is done, it will all come apart for the final finish work and paint.*

13 *The welds were then dressed with the flap disc (80-grit). These discs are nice because they don't tend to grind grooves into the metal; instead, they leave a smooth surface, which means less finish work.*

15 *To keep the metal from rusting, we sprayed it with some etching primer. This is an important step because fresh welds will flash rust almost instantly.*

16 *We finished off the notch with some semigloss black paint. Once it was painted, the notch looks pretty good.*

plenty of clearance for full-frame vehicles, and is only 1 inch deeper than the factory sump, so there is good ground clearance for lowered vehicles. This pan fits just about any application where an SBC or LS would clear. The shallow front section is just 1/4 inch deeper at the front than the factory wet-sump car LT pan.

The Drag model (part number 302-22) is the exact same oil pan, only with an added set of baffles that bolt into the pan for controlling oil slosh in extreme G situations. This is specifically designed for drag race applications. This keeps the oil in the sump surrounding the pickup where it needs to be.

At approximately $400, the Holley Retro-fit models are the cheapest solutions to the factory pan. Unless you really want to modify your frame, you can't get a pan for less. It comes with the Holley quality that you would expect, and it looks good. It is cast aluminum, which can be polished if you are looking for a fancy oil pan, painted, powder coated, or left raw. This is a simple solution for a difficult problem.

The only potential issue with this pan is that it is cast aluminum. The only real drawback is that if it is hit hard enough, it can crack, leaving the engine without oil. Is this likely? No. Not even a little. But it could happen. The factory pan is cast aluminum as well, so this really isn't much of a drawback. Because it is cast aluminum, adding additional fittings to the pan is more difficult.

Moroso Wet Sump

This Moroso fabricated pan (part number 20155) is made from sheet aluminum, which means that the pan is flat and square. There are no little bumps or raised sections in the middle of the flats that can cause clearance issues. This pan features a billet aluminum O-ring rail to match up to the block and a removable pickup tube. It uses Moroso's billet oil filter adapter, so the stock location can be used for the filter; no remote filter is required like most sheet metal pans use.

The front shallow section is 1⅞ inches deep and 14½ inches long. The rear sump measures 5⅞ inches deep and 8½ inches long, so there is plenty of clearance for most crossmembers. Included

with the pan is a trap door baffle and windage tray for better oil control.

The sheet metal construction makes it lightweight, and adding other fittings to the pan (such as an oil return for turbos) is easy. Because it is not cast, road debris is less likely to crack or split the pan because the sheet metal aluminum will dent rather than crack like cast aluminum.

The disadvantage is that the Moroso pan is nearly double the price of the Holley pan, so that is a big drawback. It might make up for it if a return fitting for turbo system needs to be added, but that can be done on both.

BRP HotRods High Clearance Sump

The High Clearance pan (part number 000-6490-00) is designed specifically to work with BRP's MuscleRods swap kits. It is a fabricated sheet metal pan like the Moroso. It features the correct O-ring placement and pickup tube. It is a very good-looking oil pan, and the aluminum is easily polished. The high-clearance version is for applications that require maximum clearance, such as Tri-Five Chevys.

BRP HotRods Extended Sump

If you have a front steer–style steering system, the extended sump pan (part number 000-6451-00) is suggested by BRP to gain the clearance without losing the 5-quart capacity. It has the advantage of sheet metal construction for less weight, and it fits BRP swap kits. However, these pans are very expensive to build, and that is reflected in the price, which is similar to the Moroso pan.

Here the issues with the factory cooler are visible. It is impossible to even get the engine into the chassis with the cooler installed, and it will hit the crossmember if it is installed afterward.

Regardless of which pan is chosen for a swap, it will likely need to be removed at some point. This is a bit different from every other GM engine that came before it. There is no oil pan gasket, instead General Motors used gray RTV silicone. Gray silicone is specifically used for high-torque applications where sensors and oil are present. It is important to use the gray silicone on the oil pan because it is the only defense against leaks.

Another potentially catastrophic issue with the oil pan install/removal is the oiling fitting O-rings. There are three: one on the pickup tube to the block and two on the oil filter ports that run to the external cooler or bypass cover. These have a habit of sticking to the engine block when removing the pan, so the pan will come off with these missing, and if you miss it during the process, you can lose them, which would mean a massive internal oil leak and no oil pressure. If you are replacing the pan or putting on a new O-ring but the old one is still on the block, you will have a really tough time getting it all to go back together.

The oil pan is installed without a gasket. This is nice because there is no need to worry about damaging it during removal. However, the factory uses a very hard silicone sealant instead. This makes it very difficult to remove the pan itself. Pry bars were used to carefully release the pan from the sealant.

We used a gasket cleaning tool, but a scraper or even a razor blade will remove the remaining sealant from the block. It must be perfectly clean.

The rubber O-ring seals tend to stick to the block when the pan is removed. Do not forget to remove the seal from the block before installing a new pan.

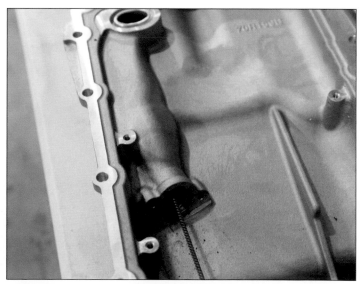

We prepped the new Holley oil pan by cleaning it well, including using bottle brushes inside every port. We used brake cleaner as a cleaning agent.

All of the fasteners for the interior pan components require medium threadlocker.

There are two versions of the Holley pan: a standard pan and a race pan with a trapdoor system to ensure that oil is always covering the pickup tube (as shown here). These components are assembled inside the pan.

The pickup tube bolts to the interior of the pan, which requires a gasket.

There can be an interference issue with the pickup tube brace if using the race baffle, which we have marked here and then ground away with a grinder. Make sure to clean the tube after this process.

The clearanced section now fits quite well with the race baffle.

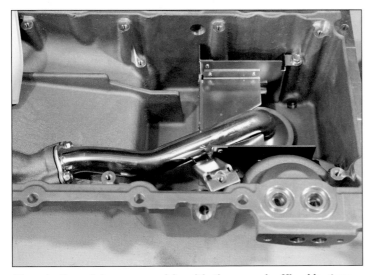

Here is the pickup assembly with the race baffle. Next up is the windage tray.

The bolts for the windage tray need threadlocker just like the others. This tray keeps the crank from whipping the oil in a frothy mess, which robs horsepower.

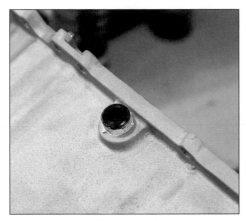

There is an optional port on the passenger's side of the pan; this is for any accessories that may require an oil drain, such as a turbo.

Don't forget to install the drain plug, which is magnetized and supplied by Holley.

If the goal is to have filtered oil, install the filter coupler, which is supplied by Holley. This piece has threadlocker already painted onto the pan threads.

Always use new O-rings for the block-to-pan seals. The Holley pan comes with these new rings.

The oil pickup port uses an oval-shaped seal; just drop it in and press it down.

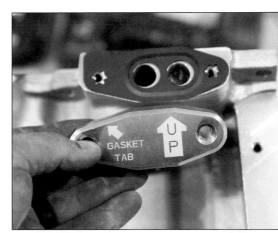

Holley provides this machined cover for the oil ports on the side of the pan, but it does not have a threaded port for a sensor.

There are a few tricks for the block seal. Run a 1/8-inch-wide bead of gray RTV silicone. Yes, it must be gray RTV because this is the correct rigid high-torque silicone for this application. It has high resistance to vibrations and all fluids, and it is sensor safe.

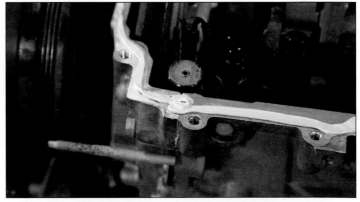

Add a little extra silicone (3/16-inch bead) over the timing-cover and rear-cover seal joints. This ensures a leak-free installation.

Two alignment dowels are on the block, and these must line up with the pan before any attempt to install the bolts.

Oil System Modifications

The LT engine is not designed for use in non-factory installations. This means that some factory components are not suitable for swaps, and some required sensors are not available on these engines from the factory. This

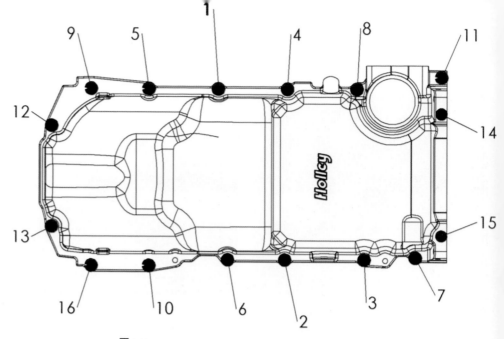

Torque:
M8 bolts to 18 lb./feet. (bolts 1 through 13).
M6 bolts to 106 lb./inch. (bolts 14 and 15).

There is a specific inside-out torque sequence for the oil pan bolts. This is very important because the cast pan could crack if not correctly torqued. This sequence is for all LT oil pans, not just the Holley pan. (Photo Courtesy Holley Performance)

includes the optional factory oil cooler, the vacuum pump, and finding a location for an oil pres-

sure sensor. Some vehicles can accept the cooler and the vacuum pump, but most do not; they are simply too large to fit in most older vehicles. Removing them presents a few challenges, but

Removing any of the factory coolers requires a bypass kit. This is the cover for the pan; it requires bolts and a metal gasket. The GM part numbers are 12630766 for the cover, 12623359 for the gasket, and 11562426 for the bolt.

There are several options for adding oil pressure sensors to the LT-series engine, one of which is the bypass cover. This port from ICT Billet is a spacer (part number 551286) that provides two ports for sensors on the bypass. This does require the factory cover. This can also be used as an oil feed/return port for a turbo system.

The metal gasket has starred holes for the bolts, which act as keepers for the bolts, making installation easier.

The orientation of the cover matters. The gasket and cover must line up with the ports in the oil pan.

If the original engine had a water cooler, there will be a large hole in the block. It will need a port plug, shown here just above the QR code pad. The plug part number is 11611351.

Chances are good that a very large Allen key is required to install the plug. We made a tool using a bolt, some washers, and a nut. This works quite well and doesn't cost anything.

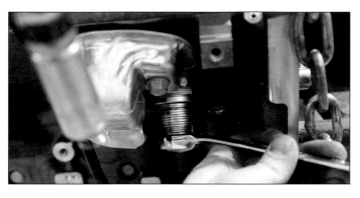

The plug needs thread sealant, so make sure to put some on the threads to avoid a coolant leak. This would be very difficult to reach with the engine installed in most cases.

these are easily rectified if they are taken care of before installing the engine.

Oil Cooler

The oil cooler is found on all 6.2L car engines with wet-sump oiling and on some truck engines. The cooler is quite interesting because it is an oil-to-water cooler. This means that the cooler is connected to the vehicle's main radiator. Coolant is pumped to the cooler, which mounts off of the driver's side of the oil pan, flows through the cooling tubes, and then back to the radiator. The oil is pushed through the cooler by the oil pump.

This unit is very large and will not clear most frame rails, much less a recirculating ball–type steering box, so it must be removed. The remaining ports can also be used for an external cooler, or they can be covered with a factory or aftermarket bypass cover. This is also the main access point for oil pressure sensors on most LT swaps.

Vacuum Pump

LT-series truck engines, specifically the L83 5.3L and L86 6.2L, use a block-mounted vacuum pump to supply the braking system with enough vacuum to operate safely. This unit sits low on the block and, (in most cases) will not clear the motor mount adapters. In some cases, it won't

even clear the chassis itself. Car engines do not use this pump.

The vacuum pump is used on truck engines to provide extra vacuum when the engine is in 4-cylinder mode. Removing the pump is required for most LT swaps, but that leaves two oiling ports on the block that must be plugged. They are 12 mm x 1.75 thread, but the GM plugs need to be used, otherwise there could be a loss of oil pressure if a standard, non-shouldered plug came out. GM part number 1546665 is the correct plug, and it even comes with thread locker on the threads. Two are required. The press port that normally feeds the pump with fresh oil can be used as a port for an oil pressure gauge; however, most motor mount adapters will interfere.

If you are considering keeping the vacuum pump, don't. GM has recently issued a recall for all LT-powered trucks due to serious safety concerns about the vacuum pump losing the ability to produce enough vacuum, leaving the trucks with a hard brake pedal.

Camaro LT1s use a different design. The AC compressor is on the driver's side, down low, and the alternator moves to the passenger's side down low. This will not work for most cars with stock recirculating ball–style steering boxes. (Photo Courtesy General Motors)

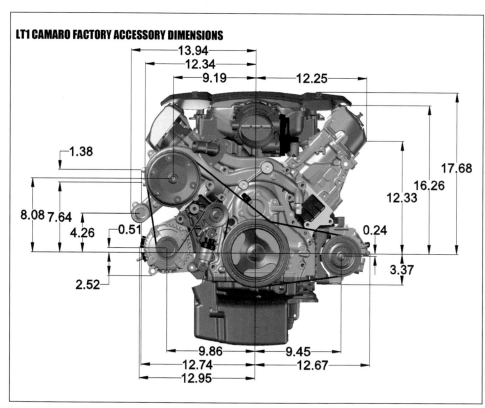

LT1 CAMARO FACTORY ACCESSORY DIMENSIONS

These measurements are for the LT1 Camaro accessory drive. (Photo Courtesy ICT Billet)

but it will be a very tight fit for front-steer muscle cars because the AC unit will likely hit the steering gearbox.

Truck Drive

The truck drive positions the water pump pulley to the driver's side of the engine. This is mostly done to provide a path for the block-mounted vacuum pump. That is the major departure from the car version.

The vacuum pump runs off a four-rib belt that is driven off

Truck drives are the reverse of the car drives because the water pump is offset to the driver's side. The same issues are present for the AC compressor, with the added complication of the vacuum pump, which should definitely be deleted. The vacuum pump is driven off a third belt on the back of the crank pulley.

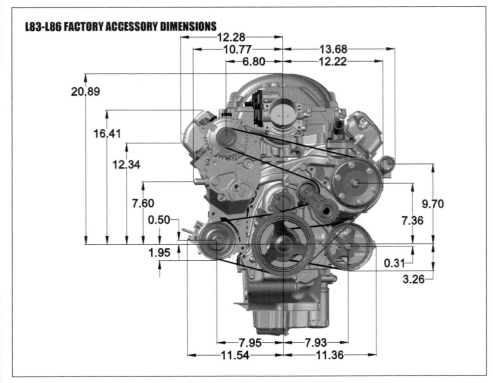

L83-L86 FACTORY ACCESSORY DIMENSIONS

These are the 2014–2018 L83/L86 accessory dimensions. (Photo Courtesy ICT Billet)

the back of the crank pulley. The crank pulley is deeper than the car pulley, similar to the LS platform. They are not interchangeable without changing the rest of the components. The alternator is mounted up high on the passenger's side, residing just outside of the valve cover. This does not present any clearance issues for most swap applications.

The vacuum pump is used to pull crankcase pressure out of the engine and to provide adequate vacuum for the power brake system. This pump is removable, and in most swaps, removal is required to clear the front crossmember. GM A-Body cars definitely require removal; however, the first-gen Camaro/Firebird can use the pump with a small dimple in the crossmember to clear the pulley. In most cases, the pump is not needed. It is possible to fabricate a bracket for a power steering pump to run off the rear belt.

For 2019, General Motors deleted the vacuum pump from all truck engines and centered the water pump, like the more traditional style. The AC compressor is still low and tight on the passenger's side. It did add an idler pulley on the driver's side, which should make it much easier to add power steering to the factory system. (Photo Courtesy General Motors)

The LT-series V-6 also uses an offset water pump to the driver's side just like the trucks. These are typically found in trucks, so there is a vacuum pump on this engine as well. (Photo Courtesy General Motors)

AC Compressor Retrofit

The AC compressor shares the same problems as the car unit: the compressor is an *always on* unit, meaning that it runs all the time, regardless of whether the AC is on or not. The compressor is controlled by the body control module (BCM). Without the BCM, the unit does not operate properly. There are swappers who have used these compressors (similar to LS units), but the problem is that the compressor will freeze up the system, because it does not cycle on and off as needed.

Even if the compressor could be used, the location does not lend itself to working in most muscle car applications. The placement is low and tight to the block, so the unit clears the frame but the fittings are positioned on the side. In GM A-Body frames, there is about 1/4 inch of clearance, so hoses cannot be attached. This placement works just fine for most trucks and wider-frame vehicles.

There is a solution to this issue without replacing the entire accessory drive. The trick is to use a small-form-factor compressor and fabricate a small adapter brace using the factory AC compressor bracket. The Sanden S7 compressor is the same diameter as the factory unit, but the position of the fittings is much better. This unit uses two pairs of ears mounted on the same plane across the compressor. This lends itself to adaptation to the factory bracket. In the correct position, the compressor provides an extra 1.5 to 2 inches of clearance compared to the factory compressor and is compatible with all aftermarket and most factory AC systems.

A little fabrication is required, but it is very simple and can be done on the car or on the bench. The bench is much easier because everything needed can be seen easily. The AC bracket is a separate piece that bolts to the engine block, so it is easily removed. The following components are needed to perform this modification:

The factory AC compressor mount does not clear most chassis designs, certainly not any GM muscle car frames. There are instances where the factory accessories must be used, such as when installing a ProCharger, so if you want AC, you need to figure this out. We designed an alternative mount that uses a Sanden SD7 compressor (the stock AC compressor is not compatible with any older AC systems).

The SD7 compressor was measured for all the mounting points. These are critical measurements, so be careful.

A piece of tubing (1/2-inch OD, 3/8-inch ID) was cut 2¼ inches long for the mounting ears.

The SD7 compressor does not fit tight enough to the factory AC bracket, so we marked where it needed to be clearanced.

Using a grinder, we cleared away to offending material. Now the compressor body fits.

Sanden S7 Compressor

- 1/2-inch OD (3/8-inch ID minimum) x 5/8-inch long spacer (can be larger OD if preferred)
- 1/2-inch square tubing
- 4 inch, 8 mm x 1¼-inch bolt (1)
- 1½ inch, 8 mm x 1¼-inch bolt (1)
- 2¼ inch, 8 mm x 1¼-inch bolt (1)
- 8 mm x 1¼ nut

The spacer and 4-inch bolt are used to mount the compressor to the lower mount on the factory bracket. There is a 1/8-inch gap on the rear ear to the factory bracket. When tightening the bolt, the sliding bolt guide will take up the space. The factory bracket is threaded, and the compressor ears are threaded. The easiest solution is to drill the bracket to remove the threads. The bracket has a long hole, so match the drill bit to the larger bore and then remove the threads. This allows the new 4-inch bolt to pass through the bracket, using the threaded ear to secure the bolt.

With the compressor bolted to the factory bracket, rotate the compressor until it contacts the bracket in the center. This will show where to trim the webbing. It does not need much; less than 1/4 inch of material needs to be removed.

The brace goes on the upper mounting ear and runs to the outside ears on the compressor. To make this brace, cut a piece 4½ inches long and then cut three sides (leaving one side intact) at 1 inch and 1¾ inches from each end, leaving 1¾ inches in the center. Make an angled cut on the opposite side of each cut, creating a wedge approximately 1/8 inch at the widest point. These cuts will be bent together and welded so the brace curves around the compressor.

The ends of the brace are also made from tubing. Round tubing or square can be used; we used square for simplicity of design and fabrication. To create the brace, a section that is 5/8 inch long is needed and another that is 1¾ inch long. The tubing we used is 1/2 inch outside, so the 5/8-inch piece is flush on one

The support brace that makes this whole thing work is 4½ inches long and made of 1/2-inch square tubing. It has to be bent in two places to follow the curve of the compressor.

side and 1/8-inch proud (protruding) on the other. The bolt head will sit on the proud side. We did this for ease of fabrication, the 1/8-inch overhang provides a little more area for welding, but it could easily be made flush as well.

Each of the cuts must be made at specific points, as shown here.

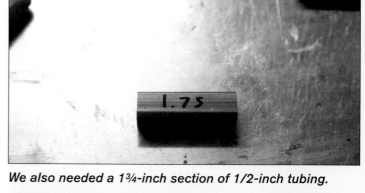

We also needed a 1¾-inch section of 1/2-inch tubing.

The final piece of tubing is a 5/8-inch section of 1/2-inch tubing.

Once the layout was confirmed, the brace was welded and smoothed out.

The pieces were bent and tack welded together for a test fit. This brace will keep the compressor in position. With this design, the compressor itself does not pivot, it is stationary.

The bracket is threaded, and we need it to be smooth bore. So, it was drilled out just enough to allow the 8-mm bolt to pass through.

We cut another 5/8-inch spacer to fill the space between the bracket and the compressor. We also used some washers as shims on the backside.

The new AC compressor assembly with the SD7 compressor and our brace is shown.

Installed on the LT1, the compressor clears the chassis by several inches and allows space for the lines, whereas the original touched the frame rail.

The longer leg is welded on the other side of the brace, opposite of the 1/8-inch overhang, running toward the rear of the engine. Position the short leg on the factory mounting tab, then with the compressor in position, bend each section together so that the long leg meets the outer ear of the compressor. Remove the brace and weld it all together. Clean up the welds and paint it to complete the modification.

As a side note, if you use Dirty Dingo sliding motor mounts, the brace may need to be lengthened slightly to clear the forward ears of the motor mount. The motor mount could also be trimmed.

Aftermarket Drives

There are several aftermarket drives available, and more will be released as swapping Gen V engines becomes more popular. Most of the systems available currently use LS-series water pumps, which simplifies the installation for most vehicles. The main draw-

There are specific crank pulleys for each accessory drive. On the left is the 2014–2018 truck pulley, and the LT1 is on the right.

The crank pulley may need to be replaced, and the entire job of swapping the accessory drive is easier with it off. Make sure to use the correct crank pulley removal tool.

back is that the accessories are mounted wide and high on the engine, just like most aftermarket LS systems. This is usually not a problem, but it can be in certain instances. The key to fitting

---

The process begins by removing all of the original drive components to get to the bare block.

The wire harness for the sensor is bolted to the block. This is removed as well.

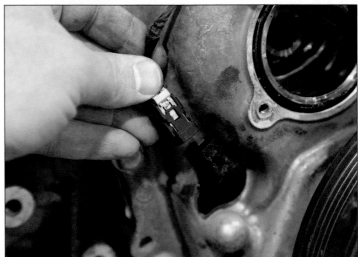

Don't forget to unplug all of the wires.

The cam sensor must be rotated, so it is removed first. Keep the fasteners.

Holley supplies a billet ring for the cam sensor, which is relocated to the proper position.

The issue is that the plug gets in the way of the water pump, but the relocator ring fixes that.

This is a common problem for most aftermarket accessory drives. This is the procedure for the Dirty Dingo drive. The sensor is spun counterclockwise, so the two holes on the passenger's side line up as shown and then the interference is marked with a pen.

Using a grinder, the sensor is trimmed so that a bolt can be installed in the third hole.

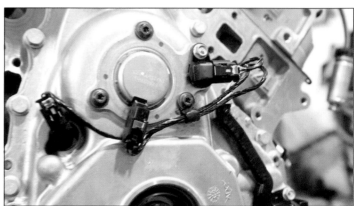

The Dirty Dingo system also requires chopping the steel wire chase and supporting the wires with a clamp as shown.

bracketry. The Dirty Dingo and ICT Billet drives are built with aluminum plates and tubing, so they are a bit more complicated to assemble.

Holley Mid-Mount

For those who need to limit the width of the engine, the Holley mid-mount system (part number 20-200) is a great option. This design completely redefines the LT accessory drive because Holley developed an all-new water pump housing that integrates all of the mounts into the pump housing. So, after you bolt on the housing and assemble the components, you are done. It is very simple, very clean, and narrow. In fact, this system is narrower than the engine itself, so this drive will fit into just about any application without issue.

The main drawback of this system is the cost. It is significantly more expensive than the standard-mount Holley system, which is not a budget piece in its own right. The benefits certainly come at a cost, but when a tight-fitting accessory drive is needed, there is simply nothing else out there that is even close. Rest assured that you are getting good quality because these Holley systems are very well designed.

Dirty Dingo

Dirty Dingo was the first company to bring a PS retrofit kit to market, and it is rather ingenious. Instead of trying to work with the offset LT water pump, the DD system replaces the LT water pump with the older LS1 Corvette center-driven pump, using a pair of billet aluminum spacers. The LT and LS share the same bolt and gasket configuration, so it is an easy swap. The spacers also provide a mounting point for a temp sender because the LT block does not have any alternate cooling ports to tap into.

The center-drive pump allows the other accessories to mount off the heads and block like a typical system. Dirty Dingo uses an aluminum plate and spacer design that fits well and looks good on the LT engine. The caveat for this is that spacing of the water pump

requires the use of the truck crank pulley.

If you have a Chevrolet Performance LT crate engine, this pulley needs to be purchased separately. The pulley is difficult to remove and install, so make sure to use the correct pulley-removal tool. A few wiring harness items must be moved as well: the cam VVT solenoid that needs to be re-clocked, along with a portion of the steel wiring harness sleeve. The alternator moves to the driver's side of the engine, so you may have to reroute or modify the wiring harness to reach the alternator.

This drive system uses either a fourth-gen Camaro type-II power steering pump with Dirty Din-

go's own pulley or a type-I Saginaw pump (different brackets). To use a type-II pump, a press-in or thread-in reservoir feed fitting is needed for the pump, which DD sells. The kit is also available with either no AC, Sanden 508-type AC compressor brackets, or GM R4-type AC brackets, which provide many options when it comes to adding AC. The factory LT series accessory brackets with AC do not clear the frame for most GM muscle cars, making this system even more desirable. Dirty Dingo provides multiple listings for belt lengths depending on the options, including alternator pulley and case size, AC type, and idler pulley location.

Moving on to the LS water pump, the Holley kit comes with new longer bolts and a set of water-pump gaskets.

Holley supplies an entirely new wiring harness.

The metal gasket should be used for the spacer to block the seal because these gaskets have the bolt retainers built in so it can be installed as an assembly.

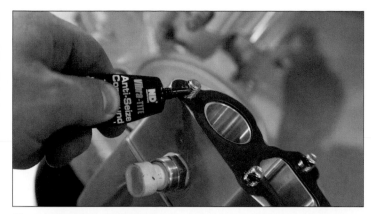

Use anti-seize on the water-pump bolts because these bolts are steel and go into an aluminum block near water.

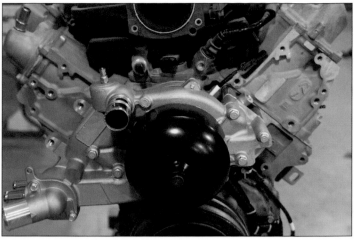

The new pump fits just like it should with the supplied spacers. General Motors switched back to this configuration for L83/L86 trucks engines in 2019.

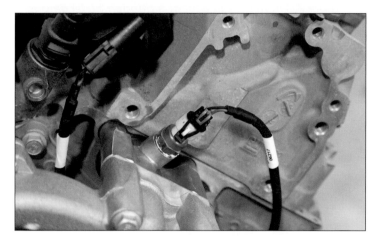

Holley's spacer provides a port for the factory water temp sensor but did not provide a port for an aftermarket sensor. Hopefully, Holley will fix this in the future.

We drilled and tapped the top of the water pump at the exit port from the pump, which is a common practice on LS swaps.

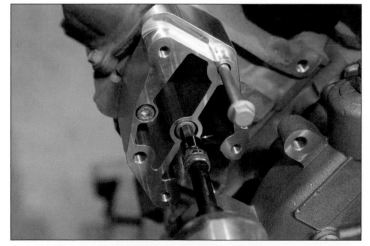

One of the really cool things about the Holley system is that the kit is designed to work on LS and LT engines, and all of the components are machined to lock together. There are a number of different length and diameter bolts for this project; make sure to use the correct ones.

These pieces here are spacers, pushing the accessory drive out to the correct location for the belt alignment. Note the wire chase for the temp sensor.

Here is how the brackets interlock with the spacers. This gives it positive placement for the brackets and ensures nothing moves around.

The upper AC mounts can accept a Sanden SD5 compressor or a GM pancake compressor, but know which type is needed before ordering it. Or just order the whole kit with the compressor.

On the driver's side is the alternator and power steering pump, which is a GM type-II unit.

Now the system is ready for the actual accessories.

The type-II pump gets an adapter bracket to mount into the main bracket on the engine. A factory GM CTS-V reservoir or an aftermarket unit can be used, but the factory tank works best.

There are two options for the idler, which is determined by whether or not a power steering pump is used. The adapter goes into whichever hole is required by the setup.

Up top, the alternator drops right into the bracket, nice and easy.

Then the idler slips over the adapter and gets bolted in.

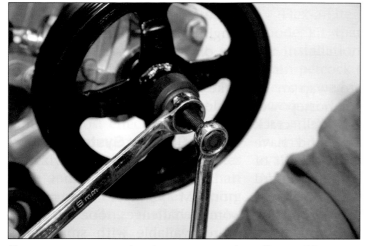

Holley supplies some spacers for the AC compressor, depending on which one is needed, so pay attention to the small parts in the package. The bolt will push the spacer on the left ear into the compressor. This is a factory-installed spacer.

If you use a power steering pump, an installer tool is needed to install the new pulley. Do this last because you don't want to remove it and reinstall it.

There are two pulleys on the upper AC bracket: one is an idler, and one is the tensioner.

Dirty Dingo's accessory drive kit looks really good. It is all machined aluminum, which provides a custom look as opposed to the factory look of the Holley system.

LT water pumps, which you may not be using if you have an aftermarket accessory drive that uses an LS-based water pump. Most of the aftermarket systems use the F-Body pump, but others use the 1997–2004 Corvette pump, making things more complicated.

LS Water Necks

The stock cast aluminum water neck points at a 90-degree angle to the right. This position works fine for many installations, but a non-stock unit may be needed to accommodate a different radiator or chassis. There are several aftermarket alternatives to the stock cast water neck. Two such options are a straight unit (which is the best unit for the early Corvettes), and a 360-degree swivel with either a 45-degree or 15-degree outlet. Each water neck must match the water pump design: 1998–2003 and 2004–up.

Since there is no mechanical fan in the way, running the upper return hose to the driver's side is a pretty simple solution if the stock radiator has a driver-side upper mount. The lower feed hose may be more difficult to cross over to the driver's side, depending on the distance between the engine and the radiator. It is possible to have the inlet and outlets moved, but the expense would likely be just as much as purchasing a new radiator.

Aftermarket Radiators

Each of the many aftermarket radiator offerings has its own benefits. There are a couple of options when it comes to ordering a radiator. The first and most simple way is to order an off-the-shelf unit

with the inlets and outlets as the manufacturer placed them. New aftermarket radiators that mount in the stock location are available for most popular cars.

You can also save some cash by purchasing a custom-fit radiator, typically sold in terms of dimension. For example, a four-core 20 x 16–inch radiator would notate a 20-inch-wide by 16-inch-tall radiator. Often, these radiators will fit in the stock location using the stock or slightly modified mounting hardware, but they can cost as much as 30 percent less.

Custom Options

The other option is to order a custom radiator. Ordering a custom radiator usually involves filling out a form and sending it in, along with a phone call or an email to discuss your needs. The ideal radiator configuration for a Gen V is to have both outlets on the passenger's side and a divider placed in the middle of the passenger tank. However, converting it to a crossover style simplifies the install.

Crossover-style tanks also ensure that the coolant takes a longer route through the tubes because all the coolant must pass through the top rows and then the bottom, doubling the surface area the coolant must pass through. Radiators built in this manner cost between $600 and $1,200, depending on the size, configuration, and manufacturer.

In addition, a transmission cooler can be run in the radiator, which keeps the transmission the same temperature year-round. This maintains a much more consistent transmission temperature

over an external transmission cooler, which allows the transmission to run cooler in the winter and warmer in the summer. Gen V engines do not tend to run hot (typically around 210°F is normal), so if your LT is running much hotter than the thermostat installed, you have a problem.

Mounting

Mounting the radiator below the engine produces ineffective cooling and excessive heat, and this is a common Gen V LT engine swap issue. This is not an issue in most muscle cars and trucks, but on many other vehicles the radiator positioning just does not allow it. In this scenario, the engines tend to hold air pockets, which leads to overheating.

There are a few solutions for bleeding air out of a cooling system. The first is to use the upper radiator hose to fill the engine. This allows the coolant to fill the engine from the top down, helping to force the air out. Once the upper hose overflows, connect it to the radiator and fill the remainder of the radiator. Fill the overflow tank to half full. Then the engine should be run with the heater running full blast and brought up to operating temperature. The overflow tank will drain into the radiator. Once the cap is removed and more coolant added, the overflow tank should remain at about one-quarter full when the engine is cool. If there is air in the system, the tank will drain; more coolant should be added until the tank remains at one-quarter full. This does not always work. It is also possible to use a purge adapter in line with the upper radiator hose. This piece contains a valve that allows the system to be purged of any remaining air.

Steam Lines

Because the Gen V LT engine shares many attributes with the Gen III/IV LS platform, you may be thinking that you need to address the steam lines. Fortunately, GM deleted this feature from the Gen V LT engine, so there are no steam lines. We include this here just to point out the fact that they don't have them. This might save you some head scratching.

Electric Fans

An electric fan is required for all Gen V LT engines. A mechanical fan can't be mounted with an offset water pump. There are many options for electric fans, both stock and aftermarket, and each fan requires custom fitting to the radiator.

For the budget-minded builder, reusing stock radiator fans is an inexpensive option. Most salvage yards give the stock fans (and maybe the radiator) for free when a complete engine is purchased. Since the Gen V platform is quite new, there will be plenty of life left in the fan motors. Of course, new fans have guarantees and can be configured to your needs.

With electric fans, correct installation is essential. If the electric fans are not installed correctly (with an electric fan shroud), they will not able to draw air through the entire radiator and will lose efficiency. The engine reaches operating temperature much faster, maximizing fuel economy because electric fans are set to run at a determined temperature.

The electric fans also operate when the engine is off, so the coolant in the radiator cools while the car is sitting. This helps keep the engine in its optimum temperature range on cruise nights, during general driving, and when parked. Purchase an electric fan that is designed for high-performance engines and has a fan shroud to maximize efficiency.

Gauge Sensors

Every engine needs to be monitored by the driver, and the LT is certainly no exception. Unlike early engines where the sensors were often mechanical and there were multiple ports for adding any sensor imaginable, the Gen V LT-series engine is monitored through the ECM. Unfortunately, the ECM is not capable of outputting that data to aftermarket gauges, and aftermarket gauges are incompatible with the signals that General Motors uses for the data anyway. This means that secondary sensors must be installed for the gauges.

General Motors did not do us any favors in this regard; there are no available water ports for the water temp, so one needs to be created. Oil pressure is not as bad, but there are only a few pathways, and the motor mounts and accessories can alter those options. We will take a look at every currently available method of getting the hard facts out of the motor and into the gauges on the dash.

Adding an oil sensor to the engine block at the vacuum pump feed is simple. A 12 mm x 1.75 adapter is needed to mate up to the gauge sending unit. A large oil pressure sensor, such as this one from Autometer, can get in the way depending on the motor mount adapters.

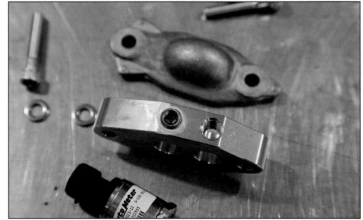

Locating the sensor on the oil bypass cover is easy and should clear on most applications. The sensor shown here is the small digital sensor from Autometer. The gauge style of the build determines which one to get.

Another option is to drill and tap the factory bypass cover, but it is pretty thin, so it may develop leaks.

Oil Ports

Pressurized oil in an LT engine can be accessed at three points: at the oil pan (similar to the LS-series engine), at the rear block drain port, and at the factory oil passage for the truck-mounted vacuum pump.

Vacuum Pump Port

LT1 engines do not have the vacuum pump. This passage is sealed off with a threaded plug, which is removable. The upper port is pressure and the lower port is the drain; do not use the lower one. This upper port is M12 x 1.75. A single brass adapter to 1/8-inch NPT is all that is needed to get the reading from here. The problem with this location is that it is directly in front of the motor mounts. Some adapter mounts, such as the Dirty Dingo sliding mounts, interfere with the port, so they can't be used. You also can't use it if you are running the vacuum pump.

Block Drain Port

There is a large plug on the rear driver's side of the block, just in front of the bellhousing flange and above the oil ports on the pan. This is a pressurized oil port. The threads are M16 x 1.5, and adapters are available to convert this port for a sensor. It may or may not be positioned well enough to use a standard barrel-type oil sensor, so elbows or a hardline adapter may be used to make the sensor fit.

Oil Pan Port

Just like all of the LS-series engines, two ports are on the side of the oil pan above the oil filter for an oil cooler. On LT1s, this is either mounted to the factory water-to-oil cooler or there is a dry-sump system, which is not very common and relatively straightforward for gaining access to the pressure ports.

Trucks, however, may have one of two options: a hardline adapter box that runs to the radiator or a bypass. The following parts are needed: 12630766 cover, 12623359 gasket, 11562426 bolt (x2), and 11611351 plug. These are GM part numbers for the bypass kit. It is easy to drill and tap the bypass for an oil pressure sensor.

If you are swapping an engine with an aftermarket oil pan, such as the Holley pan, there will already be a billet cap for the bypass, which can be used to tap for pressure. There are several of these caps available in the aftermarket specifically for this purpose. If you have the hard line cooler system and want to keep it, choose a spacer mount, which provides the port needed and allows you to keep the cooler line box. Keep in mind that most muscle cars and tight-frame cars may not clear the cooler. The car water-to-oil cooler unit definitely does not clear GM muscle cars, but it does work in GM trucks.

Water Ports

Finding a water temp port on the LT-series engine usually means drilling and tapping. Some accessory drives provide a port if the system changes the water pump, but not always. As mentioned earlier, the Holley standard mount kit provides one port for the factory sensor but not a second port.

The Dirty Dingo system provides ports for both. The factory drain plug on the driver's side of the block is a 28-mm plug, which could be used with a custom adapter for a temp sensor. The drain plug is used for the water-to-oil cooler, which is plugged off when not using that cooler. There is not currently an adapter fitting for this port, although it would not be difficult to make one. The factory plug part number is part number 11611351.

Factory Accessory Drive

If you are using a factory accessory drive, then the best option is to drill and tap the water pump housing itself. In most cases, the pump can be drilled and tapped near the factory sensor.

A standard 1/8-inch NPT sensor requires an 11/32-inch drill bit and a 1/8-inch NPT tap. Make sure there is enough room in the water port for the sensor to seat and not hit the backside. Remove the pump (best method) or stuff the port with a greased towel to catch any shavings. Vacuum out the pump afterward to be sure.

Aftermarket Drive

Many aftermarket accessory drives use LS-style water pumps that require spacers. Some kits come with a second port for a water temp sensor; some do not. Drill and tap the spacer just as described to accommodate a second sensor. You can also choose to drill and tap the top of the water pump in the same manner. The last option is to install a temp sensor in the radiator hose or in the radiator itself. These are not as accurate as engine-mounted sensors, but they are an option.

Before drilling, do some research. Follow the top port where the factory sensor is mounted across the pump. Just about any location can be drilled, but other accessories and the intake filter could interfere, so make sure that the location will work. Once the spot is found, it needs to be drilled. You may want to remove the pump from the engine to perform these steps, but it can be done on the engine if you are careful.

We located a spot and marked it with a Sharpie. This is where we will drill the water pump for the temp sensor. The temp sensor is an AutoMeter mini sensor, which is very compact. The threads are 1/8-inch NPT, so we need a 1/8-inch drill bit, an 11/32-inch drill bit, and a 1/8-inch NPT tap.

There will be a few shavings on the inside, so before leaving the pump on the engine, make sure the shavings are removed. One trick is to take off the water neck, wipe a towel with grease or heavy oil, and push it into the port behind where drilling will occur. The grease will pick up most of the shavings. The rest will be removed with a shop vac.

First, the housing is drilled with the 1/8-inch bit. This is the pilot hole so that the larger bit does not walk. The housing is rough cast, so a large bit will try to walk around quite a bit. Proceed with the full-size bit. Then use the tap to cut the required threads into the housing. Once you are done, carefully remove the towel and vacuum out any remaining shavings. If the pump was removed, blow it out with compressed air and make sure to hit all of the ports.

Installing the sensor is pretty simple, but don't forget to seal the threads with a good quality thread sealant. Teflon tape can hinder the ground connection, so stick with a liquid sealant.

The factory truck water pump temp sensor is located next to the water outlet. A second sensor can be added just below and to the right.

Be sure to use precautions. First, wipe a rag with some grease and stuff it into the water neck passage where the sensor is being added.

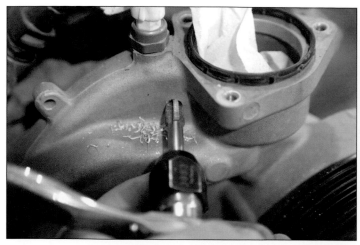

Next, drill and tap the pump housing to match the threads on the sensor. This will vary by the brand and type of sensor being used.

Most if not all of the shavings will get caught by the greasy towel. Use a shop vacuum to ensure it is all out.

It's all done, but don't forget to add some thread sealant.

The same thing can be done on top of the LS-style pump.

Liquid sealant is best for temp sensors because Teflon tape will inhibit the ground connection and alter the temp reading.

Dirty Dingo's accessory drive comes with a port in the passenger-side water-pump spacer, which is really nice and eliminates the need to drill.

LT1 ProCharger Installation

The newest GM powerhouse, the Gen V LT-series engine, is quite an evolution from the first small-block Chevy that came out in 1955. Representing the epitome of pushrod engine technology, the LT-series engine builds power quickly at a rate of efficiency that exceeds even the now-legendary LS platform, thanks in large part to the direct injection system that pushes the envelope of what is possible with gasoline.

Fuel line pressures around 2,000 psi ensure complete fuel burn at incredible levels of compression inside the engine. The stock LT1 has 11.5:1 compression, helping it generate 460 hp. Having 450 ponies is nothing to sneeze at, but what about 650? With a little squeeze from a ProCharger D1SC centrifugal supercharger, the 6.2L LT-series engine can do just that. As a quick side note, the LT1 is actually a better performance platform than the LT4 or LT5 because the LT1 has higher compression (11.5:1 versus 10.5:1), giving the LT1 a stronger base to build on for performance upgrades.

Based on the venerable ProCharger D1SC self-contained supercharger, this package is capable of adding nearly 200 hp to a 6.2L Gen V engine with just 7 psi of boost. The blower system is an easy installation, and it fits under the hood in most GM muscle car applications. There are some specific measurements to check fitment because it is on the wide side at 27 inches off the crank centerline (at 19 inches above the crank centerline). The kit is offered with or without intercooling.

The supercharger system requires the LT1 car accessory drive, and it will not work with the truck version. This means the water pump and crank pulley must be changed if it is being installed on a truck Gen V engine. It will work on a 5.3L as well, but power numbers are not available at this time. The factory accessory drive is quite compact, so it works well in most applications. The biggest hurdle is the AC compressor because it hits the frame in most GM muscle cars. It is possible to fabricate an adapter brace to allow the use of a Sanden S7 compressor (described earlier in this chapter).

Aside from adding the blower, ProCharger thought ahead for swap applications and offers a power steering pump add-on to the kit. This allows you to put power steering on an LT1 accessory drive, which is otherwise quite difficult. The pump is driven off of the secondary pulley that bolts to the factory crank pulley using a series of cam-locks. It has two sets of ribs, one for the blower, and one for the power steering pump. The cam-locks are the driving force behind the pulley, this eliminates the strain on the crank bolt. This is the system we used on our 1971 Buick GS convertible.

The engine in our project is a new GM crate engine, the Gen V LT1, which is otherwise stock. We will be swapping out the camshaft for a custom piece from Comp Cams, along with its upgraded mechanical fuel pump lobe. This will increase the mechanical pump's ability to feed the engine with enough fuel to support 750 hp. Increases beyond 750 require replacement pumps. The factory fuel injectors are

1 *Adding a ProCharger to an LT swap is a quick way to pump 40-percent more horsepower into the drivetrain. This kit is based on the LT1 accessory drive and has an option of power steering, which is nice because an aftermarket drive system can't be used.*

2 *We opted for the D1SC self-contained pressure head for the system, which increases the output. This LT1 starts out with 460 hp, but the ProCharger should push that number close to 660 at only 7 psi.*

LT1 ProCharger Installation *continued*

good up to about 1,000 hp on gasoline. The D1SC head unit is capable of 900 hp max, but if upgraded to the D1X unit, it can get to 1,000 hp. That would require upgrading the mechanical fuel pump as well as the fuel injectors themselves.

The installation is straightforward and takes just a few hours to complete. Adding the intercooler increases the power potential but also adds more time to the installation. Each vehicle has different requirements for the intercooler, so that could add a day or a week to the install time. The results are worth the effort, though, because cooler intake temps mean more potential power output.

All of the components are provided for the installation, including a new crank bolt, which is a must. The factory bolt

is a one-time-use torque-to-yield (TTY) bolt and should not be reinstalled after removal. The bracketry is well designed and fits together quite nicely for a clean finish under the hood. We opted for the black powder coat on our system.

There are a few caveats to be aware of with this blower kit. There is no tuner or tune provided, which means that the vehicle must be taken to a dyno shop after the installation. This is absolutely required and cannot be skipped. The fuel system is also not addressed in this kit, which is necessary to reach the higher range of power. Comp Cams offers camshafts with altered lobes for the mechanical fuel pump, which is the solution you will need. With the factory LT1 cam, the stock fuel system can support 750 hp. ■

3 *Our engine had been swapped into a 1971 Buick GS convertible with the Dirty Dingo kit.*

4 *The crank pulley had to go back to the Corvette LT1 pulley.*

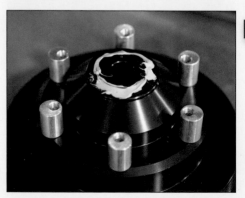

5 *ProCharger provides a bolt-on pulley that mounts to the stock crank pulley. These offset aluminum bushings align the pulley in the correct position and keep it from spinning free. A bit of gray RTV gets added to the crank pulley contact point.*

6 *The two pulleys are placed together and the offset bushings are aligned. The pulley assembly then gets installed on the crank and all the fasteners are tightened.*

7 The power steering adapter uses two plates and some spacers. We painted the kit after everything was test fit. The tensioner for the power steering belt is on the main bracket.

8 The main bracket for the compressor head is suspended off the block with a series of long spacers. These need to be preassembled because the lengths vary by location.

9 Then the main bracket simply bolts to the block.

10 Next, the compressor head is installed to the bracket. The air outlet may need to be re-clocked. The supercharger barely clears the inner fender, but it does work.

11 We mounted the supplied intercooler by building brackets that we welded to the core support.

LT1 ProCharger Installation *continued*

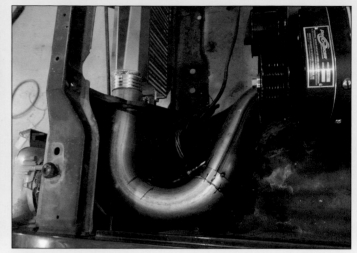

12 Next, we started laying out the tubing, which is also supplied with the kit. Note the hash marks on each joint. This ensures that you can align each section correctly when it is tack welded together. It can be tacked in the car, but that isn't always practical.

13 The tube coming from the compressor head needs a purge valve to blow out pressure when the driver lifts off the throttle. We will install that at the inside of the main bend (shown here).

14 The intercooler fits well, but to make the tubing fit, we had to have the passenger's side altered with a new air box so the exit could come out vertically at the top instead of at the side-bottom like the unit came. Side note: the unit may look crooked, but it isn't. The top and bottom of the unit are level, but the radiator itself is not. This is the because the intercooler tanks are wedge shaped.

15 We're all done. The ProCharger now feeds 7 psi of boost to the LT1 for all the tire-smoking action you can handle.

TRANSMISSIONS

The Gen V LT-series engine uses three specific transmissions from the factory: the 6-speed automatic, the 8-speed automatic, and the Tremec TR6060 6-speed manual. Thankfully, these factory offerings aren't the only options. Like the LS platform before it, the Gen V LT-series engine uses the standard SBC bolt pattern for the transmission bellhousing, allowing swappers to use just about any GM SBC-based transmission.

Automatic

It might be tempting to stick with the factory automatic transmission for a swap. Under normal circumstances, this would be a good plan. Unfortunately, there are some serious issues with both the 6- and 8-speed transmissions to know about before attempting to use one of these automatic transmissions in your swap. In one word, don't.

10L80/10L90

This is the newest transmission to be fitted to the Gen V LT-series engine, and it is a collaborative effort between Ford and General Motors. This 10-speed transmission has been used in GM vehicles since 2018, sporting 650 hp worth of power handling in the 10L90 version and 425 in the 10L80. Shifts are incredibly smooth, much better than the beleaguered 8-speed. There are three overdrive gears: 0.854, 0.689, and 0.636. This can be felt when you are doing 80 and punch the throttle to pass a car; you feel it drop a gear, pull hard, and never really feel the upshift, it is that smooth.

This transmission is a radical departure from traditional automatic architecture. There are two hydraulic pumps, neither of which is integrated into the main input shaft, which reduces the size of the transmission. These two pumps allow the vehicle to shut down the engine for the Auto-Stop feature. Because it is so new, there is not much information available for swapping, but this new unit should be a good candidate in the future.

There are many options when it comes to transmissions for LT swaps, but you should know a few things before selecting which one you want.

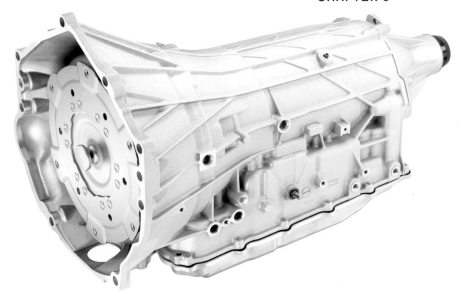

The newest GM transmission is the 10LXX series, which was codeveloped with Ford. This 10-speed transmission is very smooth when warm but very clunky in cold weather. It is very large, so fitting it under the body of most muscle cars will require heavy modification. (Photo Courtesy General Motors)

8L90

The Gen V LT engine is controlled by an ECM, and the transmission requires a separate controller. The two units talk to each other to operate both efficiently. The 8-speed 8L90 Super-Matic was a great idea, but the execution was flawed. The transmission performs so poorly that General Motors has been sued via class action over the horrendous operation of the 8-speed transmission. There are aftermarket companies that offer swap components for the 8-speed, but it is not a good idea.

General Motors has issued no less than 15 tech service bulletins (TSBs) on how to fix the inherent issues with the 8L90, but none of them work. Having personally had one of these transmissions in a 2015 GMC Crew Cab Denali, I am keenly aware of these issues. The transmission will buck wildly at low speeds, have heavy double-clunking in reverse to the point of breaking the rear differ-ential, and shudder when cruising. The 8L90 is simply a really bad idea to use in a swap. Not to mention the fact that is a very large transmission. Avoid this unit at all costs.

6L80/6L90

These transmissions are very stout and will perform well under most conditions, although the cases are quite large and do not fit in most passenger cars with a stock transmission tunnel. The 6L90 can handle up to 555 hp in passenger cars. In a truck, the weight reduces handling to 452, torque limits range from 530 to 550 ft-lbs. Both units are fully electronically controlled and can be tuned with programs such as HP Tuners.

The downside to the 6LXX transmissions is the longevity. These transmissions can last 200,000 or more miles, but they have a really bad habit of destroying themselves around 120,000 to 150,000 miles. This is due to a torque converter issue where the clutch in the converter breaks down, sending shrapnel through the transmission, eventually kill-

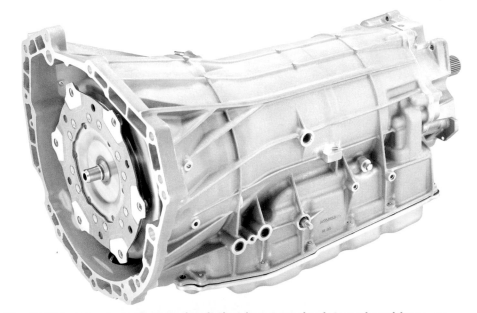

The 8LXX series is an 8-speed unit that has massive internal problems, so much so that there is a class-action lawsuit against General Motors over this transmission. Problems with this transmission include hard clunking, erratic shifting, and heavy vibrations between 50 and 75 mph. If you want a new transmission, stick with the 10LXX or 6LXX. (Photo Courtesy General Motors)

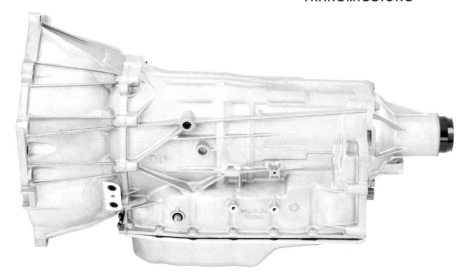

The 6LXX is a good transmission with six speeds and a smaller case than the 8LXX and 10LXX units. These are very finicky in their own right. They must be serviced every 50,000 miles, otherwise wear from the torque converter clutch can clog up the pump, which frags the entire transmission around 120,000 to 150,000 miles. It is cheaper to buy a low-mileage used 6LXX than it is to rebuild one, so keep that in mind. (Photo Courtesy General Motors)

ing the pump and the transmission. This can be mitigated by changing the fluid before the unit hits 100,000 miles and then every 50,000 miles or so.

Older GM Automatics

The 4LXX transmissions that were built for LS-platform engines will bolt directly to an LT-series engine, provided you have the correct flexplate. The LT-series flexplate uses eight bolts instead of six, so they will not swap over. This is crucial because a special LT-to-LS flexplate is needed to make the connection.

Bolting an older SBC-based transmission to an LT engine

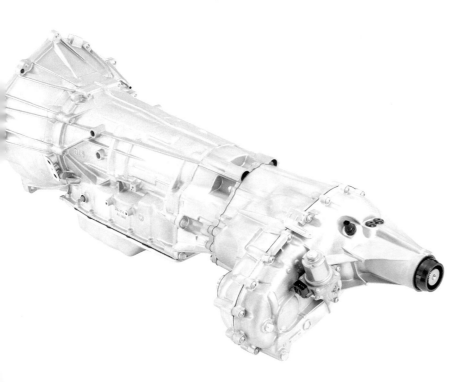

One thing that creates a few issues for swappers is that many of the engines available for swaps come out of trucks, and more trucks are 4WD than ever. The transmission for a 4WD truck won't work for a rear wheel drive (RWD) vehicle, so you may have to source a separate transmission. This is a 6L90 with 4WD transfer case attached. (Photo Courtesy General Motors)

The simplest option that fits in most applications without floor pan mods and bolts to the engine with ease is the classic 4LXX series. This is a 4-speed transmission that uses a transmission controller for tunable upgrades, and did we mention it fits? There are two main types: SBC-based and LS-based. SBC-based 4LXX transmissions have an integral bellhousing (right), while the LS-based units are longer and have a removable bellhousing. Specific flexplates and torque converters are required for each style.

requires specialized spacers and flexplates because the Gen V LT-series engine's bellhousing mounting pad is shorter. If simply bolting an old-style converter to an LT engine without a spacer, there will be major problems after only a few miles of driving. The pump seal would be ruined and begin to leak, causing the pump to fail. This is due to the fact that the converter won't center on the back

TCI makes this nice billet steel SFI-approved flexplate for the 4LXX series. They are only for the LS-based units.

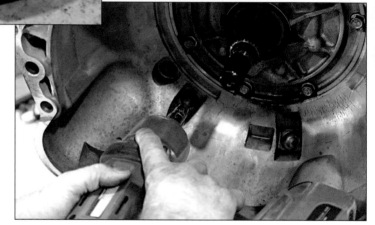

The crank of the LT engines requires an adapter ring for older automatic transmissions. This one from ICT Billet fits into the crank and provides the correct spacing and diameter for the torque converter hub.

Under the 1971 Buick GS, the bellhousing for the 4L65 has adequate clearance of the firewall and transmission tunnel. The 4LXX series case is the same as a 700R4, so it fits most GM cars.

There is a specific LT bellhousing for these transmissions, although the SBC or LS unit still works. Our crate LT1 package came with the LT bellhousing, so we removed the original from the 4L65.

On the left is the new LT bellhousing, and the LS bellhousing is on the right. The only difference we found is the additional bolt hole on the bottom left of the LT, which is actually the top left of the transmission.

We installed the new bellhousing using the original bolts and torqued them to 48 to 55 ft-lbs.

An often-overlooked spec is knowing which side of the flexplate goes to the engine side. It will be printed or etched into the steel. Do this wrong and it won't go together.

Always use new bolts for the flexplate because these are one-time-use torque-to-yield (TTY) bolts. If you upgrade to ARP reusable bolts, then no problem. These should have medium-strength threadlocker on them.

of the crank because it does not fit.

To make up for the difference, special conversion flexplates are required. Similar to the LS, the flexplate must be spaced out to accommodate the difference in depth. General Motors and many aftermarket companies offer the spacers and longer bolts over the counter. TCI has conversion flexplates for automatic transmissions and kits that include the spacer and longer bolts. You must use a crankshaft spacer, such as the Hughes Performance, TCI, or GM spacer and bolts, or a custom converter must be made with a longer crank hub.

Adapting the non-electronically controlled 700R4 and 200-4R is very difficult on an LT-series engine because there is no way to adapt the throttle valve (TV) cable. The most important component of these two overdrive transmissions is the TV cable. This crucial system tells the transmission when to shift and determines the amount of pressure sent to the clutches. If this cable is off even the slightest amount,

The torque process for the factory TTY bolts is 133 in-lbs on the first pass, then 22 ft-lbs on the second pass, and a 45-degree turn on the final pass.

the clutches will not fully engage, causing the transmission to burn up and eventually fail.

Adapting the TV cable to the drive-by-wire throttle body on an LT engine is not possible. The only option is to eliminate the TV cable altogether, which requires a constant pressure valve body. The transmission would be manual shift only without any TV cable at all. The TCI transmission constant pressure valve body removes the pressure regulation from the TV system. This valve body keeps the transmission at full-line pressure all the time, ensuring the clutches

won't burn up from slipping. An electronically controlled automatic transmission is the best bet for an LT swap.

The aftermarket also has a vast number of custom-application GM transmissions that work great for LT swaps. One in particular is the TCI 6X 6-speed automatic. This transmission is based on the 4L80E with new guts to provide six forward gears capable of handling 850 hp. These can be configured in several ways and come with a TCI transmission controller. They can even be set up for paddle shift.

Manual

The 6-speed T56 manual is the most popular manual transmission used in LT engine swaps. Formerly offered on the 1998–2002 F-Body and GTO, the T56 bellhousing, flywheel, and clutch pack are readily available. The T56 transmission fits into most GM muscle cars and requires only minor modifications, if any, to the transmission tunnel.

There are swap kits available to help make the install easier. American Powertrain offers a kit for first-generation Camaro, Firebird, and Nova. The 1968–1974 Nova requires that the transmission tunnel be enlarged. However, the clutch mechanism definitely requires some modifications. There are several ways to get around the clutch mechanism, such as to use an older-style manual clutch or a modern hydraulic clutch.

As a side note, the 1993–1997 T56s use an external clutch slave cylinder, while the later 1998-and-up units use an internal slave cylinder. The 1998-and-up T56 transmission is better suited for the Gen V LT-series engines with the correct bolt pattern and input shaft, but McLeod offers components that will adapt older T56 transmissions to the Gen V LT-series engines.

Early T56 transmissions (1993–1999 LT1-compatible units) can be converted to mate to a 2014-and-up Gen V LT-series engine with a different input shaft. The input shaft is fairly simple to change.

While the T56 is traditionally the favored 6-speed manual gearbox, Tremec has replaced it with an updated version. Actually, there are two versions: the Super Magnum T56 (the best option available in the aftermarket and from Chevrolet Performance) and the TR6060.

The Super Magnum T56 is capable of managing 700 ft-lbs of torque. The gearing is set at 2.66, 1.78, 1.3, 1.00, 0.80, and 0.63 dual overdrive. This will hit your wallet pretty hard at more than $4,500 MSRP. However, it is the strongest 6-speed manual.

The TR6060 uses essentially the original T56 case with some beefier internals. Those larger guts take up a lot of room. Because the case stayed the same size, the extra room had to be taken from somewhere. Tremec's solution was to use smaller synchronizers. The result leaves a little to be desired from the TR6060. The synchronizers are very fragile, and grinding the gears even once can wreck the synchro for that gear. Eventually, the synchro will be completely useless and the gears themselves will start burning up. Because of this, the TR6060 is not the best candidate for an LT swap. It will certainly function, but be aware that these transmissions are prone to failure.

If you select a TR6060 in your swap, make sure to use Redline D4 ATF transmission fluid or Royal Purple SynchroMax. These oils have been reported to reduce cold-shift grinding and provide better overall shifting feel for the TR6060.

There are several options for mounting the hydraulic clutch master cylinder. You can build a piecemeal kit yourself and either fabricate the mount or purchase a kit through American Powertrain, Detroit Speed, and several others. The first component is the firewall master cylinder mount. Then a line is run to the hydraulic clutch bearing. Finally, a fluid reservoir is mounted on the firewall.

On the Camaro built for this book, a hydraulic system from

The most commonly used manual transmission for any swap is the Tremec T56 6-speed. This is a Magnum T56 that had been rebuilt by American Powertrain. The other 6-speed is the TR606, which has an integral bellhousing.

American Powertrain was used. This greatly simplified the installation of the Tremec T56 6-speed into the car. The Hydramax hydraulic bearing uses a stack-up of shims to set the depth of the bearing. This allows for a perfect mesh for the clutch diaphragm. The hydraulics are relatively simple to install. The kit comes with a Wilwood master cylinder and can be bolted directly to the firewall where the stock pushrod comes through. The supplied bracket for the master cylinder is adjustable for angle.

All hydraulic release bearings (HRB) require specific spacing between the bearing and the clutch, so measure for the air gap. This is fairly easy, although

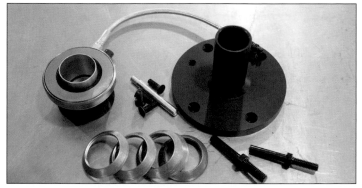

The T56 requires using a hydraulic clutch bearing. This kit is the Hydramax from American Powertrain. It is the simplest version and works like butter.

First, the base slide for the release bearing was installed onto the transmission input shaft. The silver post is the locating pin. The longest one is needed for this application.

Next, the release bearing is lubricated with some dot3 brake fluid. This keeps things sliding like they need to.

Then, we slid the bearing over the locating post and the main slide.

Now the tricky part. Using a straightedge and a set of calipers, the spacing from the bellhousing flange on the transmission to the end of the release bearing was measured. We did this three times at three different places (3, 9, and 12 o'clock) and averaged the measurements. This is measurement B.

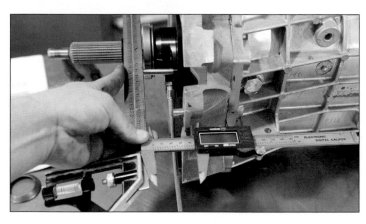

it is also really easy to get it wrong. We highly suggest doing this when the engine and transmission are out of the vehicle, because it is much easier that way.

First, install the clutch and diaphragm to the flywheel on the engine. Install the bellhousing and any spacers to the engine as well. Use a straightedge and a caliper (a measuring tape is not

We installed the American Powertrain clutch and pressure plate to the flywheel using some new ARP bolts. Don't forget to use the required clutch alignment tool. Note the gold spacer on the back of the engine block. It is required, so do not forget it.

Using the straightedge and calipers, we took three more measurements at three different positions on the bellhousing to the fingers of the diaphragm. These were averaged as well.

Next, the QuickTime bellhousing was installed to the block. This is an LS bellhousing, it is not LT specific.

Then, we did the math and determined that we needed six spacers between the bearing and the slider post. Before mating the engine and transmission together for the final time, make sure the feed line and purge line are fully installed on the bearing. It is really difficult to install it once the engine and transmission are together.

accurate enough) to measure the depth of the transmission mounting flange to the diaphragm fingers. Measure in three places (the fingers won't all be at the same depth—there is a variance), and make a note of the measurements, call it Measurement A. Next, install the HRB and carrier to the transmission, then place the straightedge across the mounting flange on the transmission and measure from the flange to the bearing surface. Make note of this measurement, which we called B.

Then you need to do math. Sorry, there's no avoiding it. Subtract Measurement B from Measurement A, then subtract an additional 0.150 inch (the air gap), and divide this by 0.090 inch (the thickness of the GM T56 spacers provided by American Powertrain). The result is how many spacers you need.

Measurement A – Measurement B – 0.150 = X / 0.090 = number of shims

A minimum of 0.1 inch and a maximum of 0.2-inch air gap is needed for the proper function of the HRB. We used six spacers on our installation with the 3/8-inch spacer plate for the L83. If you get it wrong, pull everything back out to check your work and measure several times.

If we had the right information and spacer in the beginning, this swap would have been a breeze, we wouldn't have broken our block, and we would have been done in a couple of days. The most difficult part of the entire job was actually getting the transmission to mate to the L83. Even with a transmission jack and a lift, it was quite difficult because the transmission weighs so much. That is definitely a two- to three-man job with a lift. Doing this on the floor with jack stands would be challenging, so

The T56 has several electrical connections that are important. On the passenger-side front is the reverse light switch.

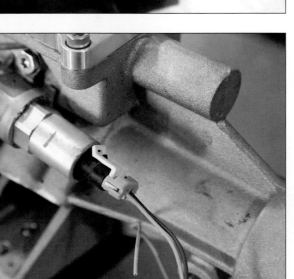

On the passenger-side rear below the shifter is the vehicle speed sensor for the electronic speedometer.

Move to the driver's side, and below the shifter is the reverse lock-out solenoid. This must be connected to make reverse shifts easier. There is a special control box from American Powertrain that makes this safe; otherwise, you could end up hitting reverse by mistake, and that would be bad.

L83 5.3-Liter Truck LT-Series Engine Crank Issue

The car-based versions of the LT-series engine are available with manual transmissions, specifically the 6.2L LT1 and LT4. This requires special machining of the crankshaft at the flywheel mounting points. For a manual transmission to properly function, a bearing must rest inside the center of the crank. For automatic transmissions, a bushing centers the torque converter.

However, with a manual, this point is crucial for proper input shaft alignment with the engine to ensure that the crank and transmission shafts are running true on the same plane. Unfortunately, the 5.3L truck LT-series engines do not have a manual-transmission option, so General Motors simply did not waste the millions of dollars it would take to machine the interior of the crank hub to accept a manual transmission. As such, bolting a manual transmission to the back of a truck LT-series engine requires some special care. We found this out the hard way when the block of a 5.3L L83 engine cracked as we torqued the bellhousing bolts.

The issue here is a 0.25-inch interference between the T56 input shaft and the L83 crank. We remedied the situation with a 3/8-inch spacer from American Powertrain. This

Mating a T56 to a 5.3L LT engine requires a 3/8-inch spacer. Without it, you get a giant split on the back of the engine block. This was a bad day.

Using the spacer requires adding longer dowels in the transmission case. This conversion to the spacer plate is a semipermanent install.

We had to use a slide hammer with a Vise-Grip welded on to get the dowel out. You can buy this or make it like we did.

issue is specific to the L83 5.3L engines and T56 manual transmissions. The TR6060 with its integrated bellhousing mate up to an L83 without this issue, so the moral of the story is to check the depth of the crank before trying to bolt the engine and transmission together.

Another side note on the L83 crank: the larger pilot shaft bearing (2010-and-up Camaro AC Delco part number CT1082) must be used because the smaller inner bearing does not fit L83 engines. If you are using an LT1 or an L86 Gen V 6.2L engine, the small bearing can be used without the need for the extra 3/8-inch spacer.

On both of our L83 engines, we had to hone the crank bore so the bearing would fit. The pilot bushing bore is too tight for the bearing as is, so we had to hone a couple thousandths using an engine bore hone. A small bottle brush hone for wheel cylinders would be an easier solution. The bearing measures 1.705 inches, but the crank bore was 1.69 inches. It is supposed to be 1.7043 to 1.7055 inches. ■

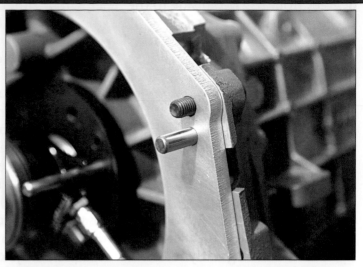

To keep the spacer lined up, we installed a few bolts in the case to secure the spacer in position.

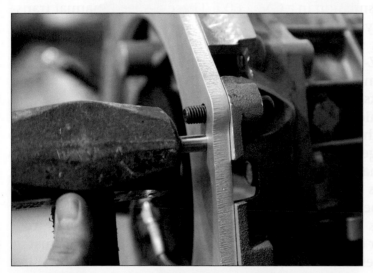

Then we drove the dowels in with a small sledgehammer. Be careful not to mushroom the dowel, but it does take some effort to get these dowels in.

The crankshaft on the L83 engine requires a little bit of honing to get the pilot bushing installed. We didn't have a small enough bottle brush hone or a brake cylinder hone, so we modified a cylinder hone and got to work. This ruined the abrasive pads for any other use, but that is okay.

be prepared for a challenge if that is your plan.

To bolt the original clutch pedal to the master cylinder pushrod, an adapter tab may be required. This tab must be fabricated and welded to the clutch pedal. The placement of this tab is absolutely critical. If the tab is too high, the pushrod will not fully engage. If it is placed too low, then the pedal will be hard to push.

Using a piece of 3/16-inch-thick mild plate steel cut to 1 x 3.055 inches, the tab should be Z'd to 0.300 inches, beginning at the 1.712-inch mark. Once bent, the tab is drilled with a 3/8-inch hole, measured from the long side of the tab at 2.629 inches on center. The placement of the tab on the pedal is 1.769 inches from the center of the square hole at the

The bearing and rod mount to the pedal assembly as shown. This will all be removed so the assembly can be installed to the firewall.

Blackjack shifter can reposition the shifter to the ideal location.

The TKO transmission does not fit quite as neatly under the body as the T56 does. The transmission fits some GM vehicles without modifications, but others (such as the very popular 1964–1972 GM A-Body platform) require a large section of the transmission tunnel be removed and replaced with a reshaped panel.

For reasons varying from nostalgia, economy, personal taste, simplicity, or originality, some builders prefer to keep the stock manual gearbox in their muscle car or truck when swapping in a Gen V LT-series engine. These swaps bolt up similar to the automatic transmissions, but they require a few specialized pieces. There are a couple of ways to do this swap as well.

The input shaft is too far from the crank with a stock bellhousing—the same situation as with the manual transmission. There are two ways to remedy this problem. The first and best way is to use a retrofit bellhousing and flywheel package. GM sells these components individually through its Chevrolet Performance dealers such as Pace Performance.

The retrofit bellhousing features thick-wall titanium-aluminum alloy construction, CNC machining including spot faced mounting holes, precision dowel-pin holes, and bores that yield a precise fit. This bellhousing bolts to all GM Gen V LT-series V-8 engines for the installation of the Muncie, T-10, Saginaw, Richmond Gear, Tremec TKO, Tremec T56-011, and other specially built transmissions. This bellhousing works with stock clutch linkage and hydraulic clutch actuators, and it includes a steel inspection cover and mounting hardware. It is designed to use a 168-tooth flywheel and standard GM starter. This bellhousing is lightweight, weighing only 15 pounds, and uses all factory linkage parts, including clutch forks, Z-bar, rubber dust boot, etc.

An original big-block Chevy manual bellhousing can also be used instead of buying a new one. The specific part needed is the 621 BBC bellhousing. This will fit most GM chassis and fit the LT engine without issue. The SBC and truck bellhousings have clearance issues. The flywheel will need to fit the LT and have a

American Powertrain's Hydramax kit uses a Wilwood master cylinder, which is very compact and easy to work with. Anything larger than this simply will not fit under the brake booster in the third-gen Camaro. The hinged bracket makes it easy to get the clutch master where it needs to go.

We made some marks on the firewall to find the correct plane for the clutch rod. The more severe the angle, the harder the pedal effort will be.

We located the original cutout in the firewall padding for the clutch rod and made a pilot hole. There is a lot of guesswork here; it is important to get really close.

The final space had to allow for the movement of the clutch rod. It was more than we had anticipated—as evidenced by the long slot top and bottom of the mounting location.

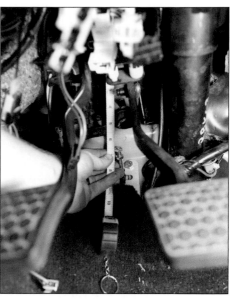

Once the master cylinder was installed, we took some measurements from the rod to the clutch pedal mounting point. This gives us the total length of the rod.

standard SBC clutch bolt pattern.

The clutch depends on which transmission is being used. GM transmissions use 10- and 26-spline input shafts, the early manuals (before 1971) typically have 10 splines, and the later units have 26 splines. That said, some other aftermarket manual transmissions have either 10 or 26 splines. The 26-spline shaft is more durable than a 10-spline shaft because it distributes the input load better. Make sure you have the right clutch for your transmission. Beyond that, any SBC clutch will work, provided the flywheel has the SBC clutch pattern.

If using the conversion flywheel, the stock-length (1.25 inches) throwout bearing works great. For stock flywheel and clutch combinations, an extended throwout bearing is required. General Motors offers a 1.75-inch length bearing (part number PT614037) for these applications. This will require using the stock mechanical pushrod clutch linkage.

Aftermarket versions of these parts in complete kit form are available from McLeod and Advance Adapters. These clutch kits are designed to adapt the Gen V LT-series engines to early-style GM manual transmissions, such as the M21/M22, SM420, SM465, and NV4500. Additionally, these kits allow the installation of Richmond Gear manual transmissions, such as the ROD 6-speed.

The Advance Adapter clutch

kits typically include a custom flywheel, 11-mm flywheel bolts, an 11-inch Centerforce pressure plate and disc, a pilot bushing spacer, throw-out bearing, collector gasket, 10-mm bellhousing bolts, 10-mm lock washers, and XRP dowel bolts. If you want to assemble your own parts, the key is to match the flywheel and the clutch to the engine while making sure the splines on the clutch disc match the transmission.

We marked and then cut the rod to length, leaving enough extra to be able to adjust the length of the rod to get the pedal in the right position.

Then we installed the rod to the master assembly and put it in the car. Don't forget the jam nuts, you do not want this to come undone while you are driving.

Early T56 Converted to Fit an LT-Series Engine

More is usually better—more power, more torque, more, more, more. When it comes to transmissions, more gears are always better. The more gears you have, the faster you go while keeping your engine in the power band. That is the key characteristic that made the Muncie M21 Rock Crusher transmission so popular. It was a close-ratio gearbox and each shift was a small jump, so it could keep the RPMs up and there was no big drop in engine speed to slow you down as you blasted down Main Street. Compared to a 3-speed, the 4s were always better. The ultimate in manual transmissions these days is the Tremec T56 6-speed.

Like the Muncies of the late 1960s, the T56 comes in a few flavors, so make sure your version matches your actual needs. Originally designed for the Viper in 1992, the T56 has been a staple of high-performance American muscle for more than 20 years. Essentially, the T56 is a 4-speed with two overdrive gears. Fourth gear is always 1:1, while fifth is 0.84 down to 0.74:1, and sixth could be as low as 0.50:1, meaning the driveshaft is spinning twice as fast as the engine.

Input shafts vary in both length and girth (spline count). This is the most important spec because matching the engine to the transmission is tricky. The older LT (1990s GM) input shaft is shorter and uses a pusher-type throwout bearing. Most late-model engine swaps, including Gen III through V GM engines, require the LS version of the input shaft. What we happened to have in the shop for our project was a LT1 T56 from a 1996 Camaro. To mate this bad boy to an LS, some modification is required. The 1991–1996 LT version of the T56 is readily available in salvage yards for just a few hundred bucks ($300 to $500 on average), so we picked one up and got to work.

We sourced an LS input shaft, an adapter plate, and a hydraulic clutch system from American Powertrain. The new front bearing for the input shaft was ordered online. Swapping the input shaft is not complicated. Simply remove the cover and carefully pull out the input shaft. Press the new bearing onto the new shaft. When installing the shaft into the transmission, a slight rotation is needed to get the gears to line up. Then reinstall the cover.

The next part requires a dial indicator, a magnetic base, a piece of steel, and some clamps. Place the steel across the front cover, clamp it down, and lock the dial indicator to rest against the end of the input shaft. The factory spec is 0.0005 inch to 0.0035 inch. Push the input shaft in, zero the indicator, and then pull the shaft out. Note this number.

The end play is adjustable through the shims behind the race in the front cover. Once the end play is set, apply some assembly lube to the front bearing, install the front cover with a carefully placed thin bead of silicone, and torque to 26 ft-lbs. Too much silicone can gum up the internals of the transmission, so use just enough to seal the case.

With these steps handled, the T56 is ready for installation into any 2014-and-up LT-series swap. Seeing as how the early versions of the T56 are so much cheaper than the newer LS version, it just makes sense. ■

Begin by removing the bolts for the front cover. Keep these bolts because they will be reused. Make sure to drain the transmission first.

Using a mallet, lightly tap the corners of the front cover to separate it from the case. It may be glued in place with silicone.

Next, the input shaft assembly is carefully removed from the transmission. A slight twist will walk the shaft out.

The LT1 (GM Gen II engine) shaft is on the left, and the new LS-style shaft is on the right. Notice how it is about an inch longer. The two are otherwise identical.

A new inner bearing race is also needed, which is also pressed into the back of the gear.

The cover is sealed with silicone, but it only needs a thin layer. Too much can cause problems with the internals.

LT1 T56s use a clutch fork with an external slave cylinder. The stub on the transmission has to go. We chopped it off with a cut-off wheel.

To set the end play, the shaft and front cover are reinstalled and a dial indicator is used to check the end play. The factory tolerance is 0.0005 to 0.0035 inch. Ours was loose, so it needs more shims.

Before the final cover install, pre-lube the bearing with some assembly lube. Don't forget this step.

Using a hydraulic press and a length of thick-wall tubing, the new bearing was pressed onto the new input shaft.

The shims do not come with the bearings, but we found some extra shims from a rear differential kit that fit just fine. We replaced the race with a new one and pressed it into the front cover.

Make sure the guide pins line up with the case and then install the original bolts and torque them to 25 ft-lbs.

WIRING

The most daunting task of any fuel-injected engine swap is the wiring. While these engines are very complicated on the wiring front, the reality is that it just isn't that big of a deal these days. If an original harness that is not cut up can be found (it's harder than you might think), it can be converted, but the best bet is to simply purchase a new harness where all the hard parts are done for you. Add in the options on fuel systems, transmissions, and tuning, and you can quickly find yourself with a pile of wires and without much guidance.

The LT engine harness provides all of the necessary wires for the engine and fuel system, but unlike other engine series, the Gen V LT-series ECM does not control the transmission. So, none of those wires are in the harness. This makes the factory harness better suited to be used as is without removing unnecessary wires. There simply are not many wires that won't be needed.

Factory Harness

If the plan is to use a stock harness, you will need to be able to sort out the power, ground, and ignition wires. The following table is a complete pin-out for a 2014 L83 5.3L Gen V engine harness. General Motors does make some minor changes here and there, so be careful when using this information on other-year engines.

All Gen V LT-series engines use electric cooling fans and dead-head fuel pumps that are pulse-width modulated (PWM). This means that the signal coming from these wires must be used with PWM-capable fans and pumps. They *cannot* just be wired to a regular fan or pump and expected to work. They will not work, and it will cause serious damage.

Another important note is that the fuel system module must remain with the engine, and it must be properly connected. There is more on the fuel system in chapter 8, but it is mentioned here because you must not delete the fuel pump module wiring. The fuel pressure signal is required for the ECM to properly control the engine.

The wiring harness calls for a brake pedal pressure switch, but on a swap, all that is needed is to wire pin 57 white/d-blu (blue X1 plug) to the brake light switch. Gen V ECMs need to see +12 volts with the brakes applied, which is opposite of the Gen III/IV engines (which require a ground trigger). The brake pedal pressure sensor is not needed.

Wiring can be very frustrating, but a premade harness like this one from Chevrolet Performance makes this otherwise daunting task easy. All the terminations are made; it even comes with an OEM fuel block.

Pin	Color	Function
K20 Engine Control Module X1 (Blue)		
1	Not occupied	—
2	D-BU/WH	Fuel line pressure sensor signal
3	Not occupied	—
4	YE/WH	Throttle inlet absolute pressure sensor signal
5	WH/RD	Throttle inlet absolute pressure sensor 5-volt reference
6	L-GN	AC refrigerant pressure sensor signal
7	Not occupied	—
8	BK/YE	Fuel line pressure sensor low reference
9	D-BU/WH	Fuel Tank Pressure Sensor Signal
10	YE/RD	Fuel tank pressure sensor 5-volt reference
11	Not occupied	—
12	Not occupied	—
13	D-BU/GY	Outside ambient air temperature sensor signal
14	WH/RD	Accelerator pedal position 5-volt reference (1)
15	YE/WH	Accelerator pedal position signal (1)
16	Not occupied	—
17	Not occupied	—
18	Not occupied	—
19	Not occupied	—
20	Not occupied	—
21	BN/RD	AC pressure sensor 5-volt reference
22	BK/BN	AC refrigerant pressure sensor low reference
23	Not occupied	—
24	BN/RD	Fuel line pressure sensor 5-volt reference
25	D-BUNT	Primary fuel level sensor signal
26	BK/L-GN	Fuel level sensor low reference
27	Not occupied	—
28	Not occupied	—
29	Not occupied	—
30	BK/D-BU	Accelerator pedal position low reference (1)
31	Not occupied	—
32	WH/GY	AC compressor clutch relay control
33	BN/RD	Accelerator pedal position 5-volt reference (2)
34	L-GN/WH	Accelerator pedal position signal (2)
35	Not occupied	—
36	D-BU/BK	High-speed GMLAN serial data (+) (3)
37	WH 37	High-speed GMLAN serial data (-) (3)
38	WH	Fuel temperature/composition signal
39	D-BU	High-speed GMLAN serial data (+) (1)
40	WH	High-speed GMLAN serial data (-) (1)
41	Not occupied	—
42	Not occupied	—
43	Not occupied	—

Pin	Color	Function
K20 Engine Control Module X1 (Blue)		
44	GY	Fuel pump controller data out signal
45	Not occupied	—
46	BN/WH	Check engine indicator control
47	WH	Brake apply sensor supply voltage (location has two wires)
47	GY/RD	Clutch apply sensor voltage reference
48	D-BU/YE	Brake apply sensor signal (location has two wires)
48	YE	Clutch apply sensor signal
49	Not occupied	—
50	Not occupied	—
51	VT/L-GN	Run/crank ignition 1 voltage
52	RD/BN	Battery positive voltage
53	BKPU	Accelerator pedal position low reference (2)
54	Not occupied	—
55	Not occupied	—
56	Not occupied	—
57	WH/D-BU	Cruise/ETC/TCC brake signal
58	Not occupied	—
59	BN/YE	High-speed cooling fan relay control
60	YE/WH	4WD wheel lock indicator
61	GY/BK	4WD low signal
62	VT/D-BU	Powertrain main relay fused supply (2)
63	YE/BK	Starter enable relay control
64	Not occupied	—
65	Not occupied	—
66	WH	EVAP canister vent solenoid control
67	VT/D-BU	Powertrain main relay fused supply (3)
68	BK/BN	Brake apply sensor low reference (location has two wires)
68	BK/GY	Clutch apply sensor low reference
69	Not occupied	—
70	VT/YE	Accessory wakeup serial data
71	Not occupied	—
72	YE	Powertrain relay coil control
73	VT/D-BU	Powertrain main relay fused supply (1)

Pin	Color	Function
K20 Engine Control Module X2 (Black)		
1	Not occupied	—
2	Not occupied	—
3	BK/L-GN	Fuel rail pressure sensor low reference
4	Not occupied	—
5	GY/BK	Output speed (digital) 5-volt sensor reference
6	VT/WH	Vehicle speed sensor signal
7	BKL-GN	Vehicle speed sensor low reference
8	L-GN	Output speed (digital) signal (location has two wires)
8	BN/WH	Output speed high (replicated TOS) input signal

Aftermarket Harnesses

The nicest part of aftermarket harnesses is that they are customizable. The factory engine harnesses are not very long, so install the ECM and other components wherever you can. This always means that the ECM will be under the hood, there is no hiding it, and you are limited in terms of where sensors are located on the engine. This is okay as long as the donor engine and harness are from the same vehicle. However, if you need to convert the engine to a different accessory drive, then certain items will need to be relocated.

That may mean extending wires, which can be tricky on low-voltage sensor wiring. An aftermarket harness can be ordered for whatever style engine you have. It can make adjustments for length, so if an extra 6 feet is needed for the ECM terminals, that is possible. The other benefits include getting the wires needed (and none of the ones you don't), that it is all new wiring, and that the wires are

well-labeled. Factory harnesses are not labeled at all, and it can be really tricky sorting out all of the plugs, especially when many of the plugs are the same.

Several companies offer harnesses for Gen V engines, including Chevrolet Performance, Current Performance, Howell, and Speartech. For the most part, they are all similar. The key is finding the functions that you need at the price that you want. The most inexpensive example is $800, and most are near that price, but they can run into the $2,000 range with options such as tuned ECMs, sensor kits, and transmission controls. A custom harness usually takes a few weeks to receive.

Some companies delete the fuel system components from their harnesses. This requires using a standard electronic fuel pump and regulator. This creates a very serious runtime issue for Gen V engines, particularly at start-up, idle, and wide-open throttle (WOT). The problem is that the ECM must know the precise pressure in the fuel system

so that it can compensate on the injector side.

All Gen V LT-series engines are direct injected, so the fuel pressure is absolutely critical. A non-PWM fuel system can be used, but the ECM must have the fuel pressure sensor installed in the fuel line, and that requires the fuel control module. At this point, only the right pump is needed to operate the engine. There is absolutely no good reason to not use the correct PWM-controlled fuel pump with a Gen V LT-series engine.

The only loose wires to deal with are battery voltage, ignition, and grounds. The rest of the wires are pre-terminated and plug directly into each component. Even the starter on Gen V engines uses a plug instead of wire studs. The main battery wire for the starter is a threaded stud, however. All of the terminals on the harness are Metri-Pack style terminals with small silicone O-rings, nylon molex plugs, and small crimp-style pins. If one of these needs to be replaced, a special crimping tool is needed.

The Chevrolet Performance harness kit looks OEM because it is. This kit is the simplest to install of all the harness kits. It is expensive, but you also know that everything is 100-percent correct.

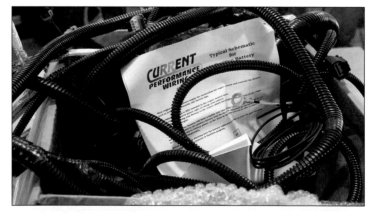

For the Camaro project, we chose a Current Performance wiring harness. They made sure it had all of the correct fuel system controls intact, which is critical for correct operation of the LT engine.

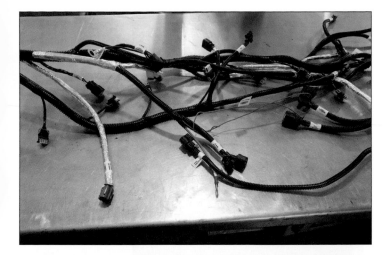

The critical part of any LT harness is the inclusion of the entire fuel controller harness. Some harness makers delete the fuel control, which is just a bad idea. Using the factory fuel design is the only way to get optimum performance.

There are some real problem areas on the LT engines, most notable is the location of the fuel injector harness terminals, which is right on the back of the engine. In fact, they overhang the flexplate. On some vehicles, it may need to make the connection with the engine lifted, it is that big. On the Camaro, we bent them, and it was still really close to the firewall.

The Current Performance harness was well laid out, fully loomed, and grouped correctly for the engine we were using. Different versions of the LT engine have a few variations for sensor placement, so the harness has to match; otherwise, you will be making changes.

Each harness must have certain fuses, and Current Performance uses a very small fuse box. This makes hiding the fuses away relatively easy. Some harnesses don't provide a box at all, just individual fuse holders.

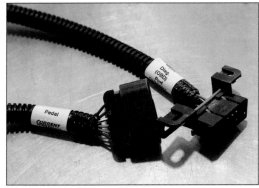

All LTs use an electronic throttle and need the OBDII port to access the ECM. The port needs to be readily accessible.

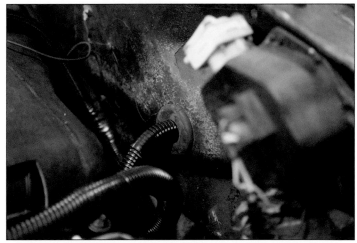

Remember that any time the harness goes through metal, it must be protected with a grommet. The plastic loom is not good enough.

On the Camaro, we ran the harness through the passenger fender, matching the original harness.

There are a handful of loose wires that must come into the interior of the car in addition to the ECM plugs.

We laid the rest of the harness out over the engine so that we could figure out the best routing. This part requires some careful thinking because servicing the engine must be taken into account.

The main engine harness was looped under the hoses, but the other half of the Y in the loom was the fuse box. So, it was routed to the passenger-side fender.

We routed the harness under the heater hoses. See how tight the harness fits to the firewall; there just isn't much room.

Sensors

When it comes to monitoring a Gen V engine, there are two ways to do it. Either run aftermarket gauges and sensors or pull the data from the ECM. Adding sensors to a Gen V LT-series engine is not difficult with one exception: the tachometer.

There is no usable tach signal coming from the engine or the ECM. Instead of a tach signal, there is a camshaft signal on the black X2 connector, pin 39 (white wire). Because it is a camshaft sensor reading, it may not be able to be read by the tach,

which has to be set to 4-cylinder and requires a pull-up resistor just like LS engines. Many LT ECMs are not programed for this signal, and the wire may not even be in that terminal on the wire harness. This makes adding a tach much more difficult. When using this signal, the tach may act erratic at idle due to the fact that the signal is coming off a cam sensor.

If the harness is not wired for the cam signal, it can be added to the molex plug in the harness. Once the wire is installed, you will also need to activate the signal in the ECM. This requires programming with HP Tuners or

similar full-access tuning software. A handheld unit will not get deep enough into the ECM to activate this signal.

One of the most unique aspects of the Chevrolet Performance swap harness and ECM is that this kit comes with the tach wire. The ECM is programmed to provide a standard tach signal. This is something that only General Motors can do because engineers can create whatever they want inside the ECM, whereas aftermarket tuners cannot alter the actual code of the ECM. This is one major benefit of the Chevrolet Performance controller kit.

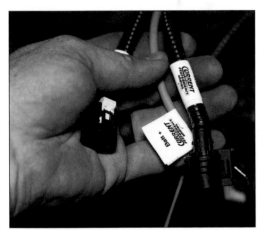

We ran the lower engine connections under the car along the rear of the block, away from the exhaust. The starter connections are all plugs with the exception of the battery lug. It just doesn't get much easier.

One thing that absolutely cannot be overlooked is the number of grounds on the LT engines. Not even one of them can be deleted; they must go where they are labeled. Otherwise, you can end up with some funky ground loops and the engine may not run right.

There is little room, so the heater hoses were run between the intake and the coil harness. This is due to the firewall exit of the hoses and the height of an LT engine.

Everything is plug and play. There is very little guesswork on the wires themselves, but there are no connection instructions, just the harness and a note on battery connections.

Working with Metri-Pack Terminals

Metri-Pack (MP) is the newest generation of modular terminals. These terminals use flat blades and rectangular female terminals in a metric size, which is where *Metri* comes from. Most of the MP terminals are sealed, but not all of them are sealed because interior connectors don't need to be sealed. The biggest improvement in modular termination with the MP design is the connector block itself. There are tons of shapes and styles, with inline, double stack, and more variations in different series. The 150 series uses 1.5-mm terminals and is rated at 14 amps, the 280 series uses 2.8-mm terminals and services up to 30 amps, the 480 series has a (are you ready for this) 4.8-mm terminal rated at 42 amps, and the 630 series services 46 amps with, drum roll please, 6.3-mm terminals.

Assembling these types of terminals requires a couple of special tools. The main tool is a pair of barrel-style crimping tools, preferably with ratcheting action that does not unlock the tool until the crimp is completed. The ratcheting tools are fairly expensive, but there are non-ratcheting tools available. The second tool is a small-diameter pick used to disassemble the terminals from the block.

The key to crimping MP terminals is to use the right size terminal for the wire and not to under/over-crimp the terminal. Under-crimping will result in a loose-fitting connection, and over-crimping can actually cut the terminal in half.

Removing terminal pins is fairly simple. A special release tool can be purchased for MP terminals, but a small flat-blade screwdriver or pick works as well. Slip the tool into the front side of the terminal to release the pin and pull off the wire from the back side. Some terminals release from the back of the block. You should be able to see the pin inside the terminal block.

Begin the crimp by stripping a small length of wire. How much depends on the terminal and the size of the wire (typically 3/16 to 1/4 inch). Place the wire inside the crimping lugs on the terminal. If the terminal is insulated, be sure to slide the silicone plug over the wire before the crimp terminal.

Here is where it gets tricky. The tool has a special shape that forms an *M* on the tangs that pushes into the wire and clamps it to the terminals. Sometimes it is easier to make this happen by gently squeezing the tangs inward a tad before loading the terminal into the tool. Once set, squeeze the crimp tool until the ratchet releases.

For insulated terminals, slide the silicone plug up to the terminal and into the insulator lugs (the rearmost large lugs), and crimp these with the corresponding round die on the crimping tool.

With all of the terminations complete, the terminals are slid into the terminal block until the pin clicks into place. You should feel it click, and you might hear it too. Give the wire a tug to make sure it is set. ∎

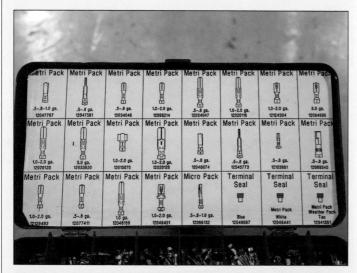

Metri-Pack terminals come in different sizes for different gauges of wire. It isn't necessary to have a selection like this, but it is nice to have them on hand when you need one. They are quite hard to find locally.

A special crimp tool is required for these terminals. A standard crimper cannot be used. These crimpers apply the exact correct pressure and automatically release when the crimp is complete.

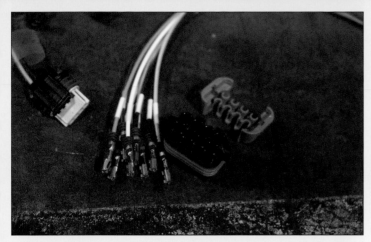

Metri-Pack terminals are made by crimping the wires and then installing the terminals into the correct location of a molex plug.

The components of Metri-Pack terminals are the wire, the terminal end, and a silicone boot. Sometimes it is possible to leave out the boot, but if the plug can see any moisture, it should be booted.

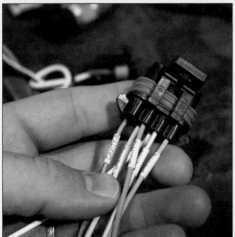

The molex plugs are hard to find, so be really careful with them. Breaking one can be difficult to replace. If the donor engine has the molex plugs from the original harness still on the sensors, it is a good idea to save

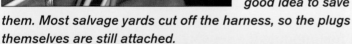

them. Most salvage yards cut off the harness, so the plugs themselves are still attached.

There are some unique terminal types, so when you are making a Metri-Pack plug, make sure the terminals match.

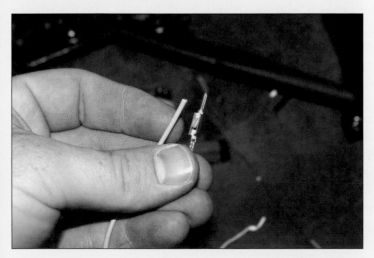

It is possible to over-crimp the terminal if it rotates in the tool as it is compressed. This one just broke off at the wire.

Truck intakes put the throttle body connection on the passenger's side, but we swapped the intake from an LT1 onto our L83 to get a lower profile for hood clearance. This put the throttle body connector on the driver's side. Oops.

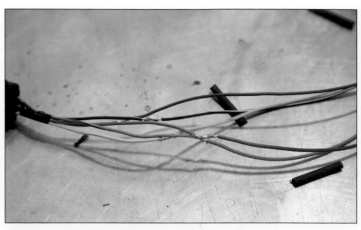

All is not lost. We simply cut the loom apart and extended each wire, using scrap wire to match the colors of each wire. Soldering is required; don't use crimps for these critical low-voltage connections.

Under the car, there is even less room. Be really careful with routing to the driver's side of the engine.

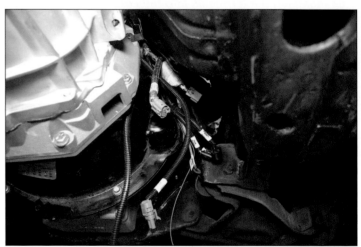

There are even more wires on the passenger's side.

If you use anything other than the Chevrolet Performance controller kit, the solution comes in the form of a Dakota Digital BIM module. This module connects directly to the OBDII port, and it pulls all of the pertinent information from the ECM, including oil pressure, water temp, speed (when wired to the ECM from the trans), and the tach signal. It is a plug-and-play unit, which really makes wiring easy. The only thing not included is the fuel level, which can be pulled from a factory GM fuel module if you use one in the tank. It makes installing custom gauges a breeze. Dakota Digital makes different BIM modules to operate its digital dash clusters or traditional gauges too.

The signals coming from the BIM module can be used to drive most aftermarket gauges. Otherwise, Dakota Digital gauges can be used, which are available in direct-fit or universal applications. What makes this nice is that there is no need to run extra sensors and adapters. The BIM module pulls the data directly from the ECM via the OBDII data port. To access the data port, simply remove the BIM plug and plug in the scan tool or programmer.

There are other options for pulling the tach signal from the ECM, including a Lingenfelter Performance CAN2-002 unit. It is similar to the BIM module, except that this one can be configured to operate aftermarket or factory gauges as long as they use analog signals.

There is one other possible solution for a tach signal, which is to wire the engine with the AutoMeter 9117 module. This module wires in line with the ignition coils and creates a tach signal from the power supply of each coil. Wiring this module into the harness of the engine will take a little bit of time and is a complicated amount of wiring, but it is a reliable method of generating the required tach signal for an aftermarket tach.

Outside of the tach signal, adding sensors for aftermarket gauges is not difficult, but there will need to be some modifications. There are a few points available for oil pressure, which depend on which accessory drive, oiling system, and motor mount adapters are being used.

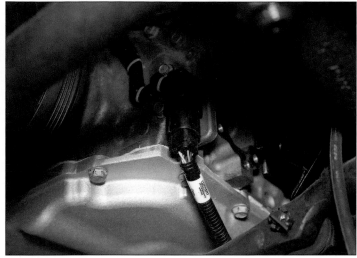

Some of the under-engine wires are for sensors on the engine, such as the cam sensor harness. Others are for the fuel system, oxygen, and transmission sensors.

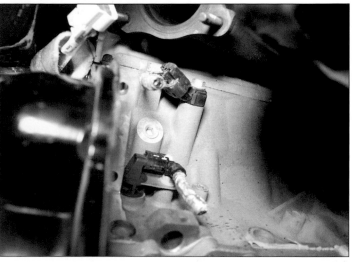

There is a knock sensor and crank position sensor on the lower passenger's side.

The leads for the oxygen sensors cannot be lengthened, but extensions can be installed if longer wires are needed. Just don't cut and try to extend the wires on the oxygen sensor. The signals from the oxygen sensors are made through high-impedance load voltage, and they use the wires themselves for cold air referencing. Modifying these wires is very bad.

Because this car is a 1987, it has an OBDI connector, which matched up to the OBDII connector. So, the location will stay the same.

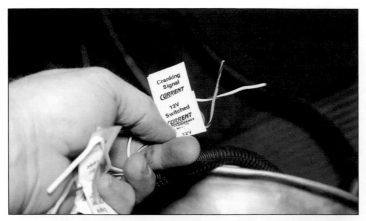

There are only a few non-terminated wires in the harness, and these two are for the ignition switch.

The pedal plug must match the pedal, which works best with the CTS pedal (part number 10379038).

This pedal fits in most applications, and the mount is very basic. So, it can be cut up and made to do whatever is needed.

Dakota Digital Gauges Installation

Swapping an LT into a third-gen Camaro brings it into the 21st century, but the factory gauges certainly leave a lot to be desired. Aftermarket gauges are not difficult to install into the factory gauge panel, but that doesn't always provide the stealth look, which is what we were after. To solve this issue, we went with Dakota Digital's direct-fit VHX series of gauges, which have traditional needles and look amazing in the dash. Combined with the BIM module, the entire system can be installed in just a couple of hours with minimal wiring. This gauge setup can be used with the original engine too; it does not have to be a swap.

The installation is very straightforward. A wiring diagram is needed for the factory gauge cluster but does not come with the kit. So, you will need some basic internet skills to find what you need, depending on what you want to connect. The control module can read the factory warning lights, as well as turn signal, high beam, cruise, etc. All of the trigger wires are in the two factory terminals.

We selected the wires we wanted to use (turn signals, high beam, ignition 12V, and fuel level), cut them from the plug, and then taped up the terminals for safety. The terminals could be removed altogether, but it is not necessary. For a 1987 Camaro, the wires needed are as follows:

Passenger-side terminal: Blue is right-hand turn signal, pink/black strip is ignition 12V, pink is fuel level

Driver-side terminal: Blue is left-hand turn signal, green is high-beam indicator

Everything else we need comes from the BIM module, which we secured under the steering column trim cover. If we need to access the OBDII port for tuning or checking codes, we simply disconnect the BIM module plug. Installing the Dakota Digital gauge cluster is likely the simplest gauge install possible; it really is that fast and easy. The only wires we can't get from the factory terminals are ground (which are available in the terminal, but we didn't trust the wire size) and battery power (which we pulled from the fuse box). ■

1 The Camaro project needed up-to-date gauges, but the original looks were important to keeping the vibe of the interior. A set of Dakota Digital gauges solved all the issues, including pulling sensor data from the LT ECM.

2 The factory gauges are okay. They certainly fit the malaise-era of automobile design, but they are not capable of reading the signals from the new LT-series ECM, so they have to go.

3 The trim bezel on the dash is retained by several T15 Torx screws. They seem like they are fake, but they are real. Keep the screws, they will be needed later.

4 The bezel comes off and exposes the cluster itself. There are two hidden 10-mm fasteners inside the warning light cavities. Remove these and the cluster is free.

5 To get the cluster out, lower the steering column. To do this, loosen the two nuts under the dash until the column drops enough to allow the cluster to come out. It isn't necessary to drop the column all of the way—just lower it.

Dakota Digital Gauges Installation *continued*

6 We cleaned the bezel and removed the factory warning light lenses on the bottom. The Dakota Digital kit comes with replacement covers. We used super glue to secure them in place.

7 The BIM module attaches to the OBDII port on the bottom of the dash. We previously installed the new LT ECM port in the same place as the factory port.

8 Then we secured the module to the dash behind the column trim plate. This will need to be accessed for programming.

9 Before moving to the wiring, we used some tape to mark off the wiring ports we won't be using for the install. This is not necessary, but it keeps things easy when you are in the dash. These gauges are capable of displaying nearly all the info you get from a factory dash.

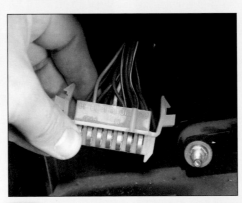

10 There are two terminals behind the cluster: one on each side. You will need to pop it out to access the wires behind it.

11 We cut the wires we needed and connected them to the new wires that we are using for the control module. This is the passenger-side terminal, where blue is the right-hand turn signal, the pink/black strip is the ignition 12V, and pink is the fuel level. On the driver-side plug, the wires we used were: blue for the left-hand turn signal and green for the high-beam indicator.

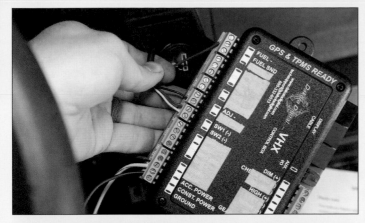

12 Each wire was routed to the control module, trimmed, and installed into the terminal rails. The module used reliable clamping blocks, which was nice.

13 The BIM module pin and the Cat5 cables for the gauge cluster were also plugged in and the module was mounted with double-sided tape behind the cluster. There is about 1/4 inch of clearance once the new cluster is mounted, so it is about perfect.

14 Two plugs connect to the cluster itself: the turn signals and the Cat5 cable.

15 The cluster is only held in by two screws: the two upper centermost screws. Drop the cluster in and align it with those holes. The original bezel then drops in over the cluster.

16 We installed the original fasteners into the bezel and the cluster. This secures everything so we can finish installing the rest of the bezel fasteners.

17 All done. We need to program the control and BIM modules, which varies by application and needs but only takes an hour or so. The new gauges look great, now we need to do something about that steering wheel.

ECMs AND CONTROLLERS

Gen V LT-series engines require an ECM. There is no carbureted version, so every swap will require a fully functional ECM to control it. Unlike the LS platform, the LTs do not have any aftermarket ECM options, at least not yet. There are some other changes inside the Gen V ECMs as well that are critically important.

The ECMs can be found relatively cheap. It is a good idea to purchase the ECM with the engine if you are buying a take-out engine. When buying a crate engine, know what year it is based on.

ECMs

There are two versions of the E92 (GM's code for the ECM) controller, and this is based on the year of the engine. In 2017, General Motors changed the ECM and harness, so there is a 2014–2016 ECM and harness, and a 2017–2018 ECM and harness, and now 2019-up ECMs. The ECM must match the harness and sensors on the engine.

It is important to make a note of the VIN for the ECM being used. This is required for tuning; the ECM can't be accessed without it. The ECM measures $9\frac{7}{8}$ inches long x $7\frac{1}{8}$ inches wide x $1\frac{5}{8}$ inches deep with a 1/4-inch inset lip for mounting. There are no other provisions for mounting. The case cannot be drilled at any point, so it must be mounted in a clamshell or with straps.

The factory ECM mounts are very large and can be used in trucks, but most muscle cars don't have the space for the factory-style mount. Dirty Dingo offers aluminum mounts, or you could make your own. The ECM case is weatherproof, as long as water does not collect and submerge any part of the unit.

Building a clamshell is a simple process. The cutout for the ECM should be $6\frac{1}{2}$ inches x $9\frac{1}{4}$ inches. The thickness of the ECM at the flange is 1/2 inch (to the back of the case). There is a small lip on the back of the case, so the options are to make two rings and sandwich the 1/4-inch lip or make a solid base and use one ring on the front side of the ECM. Keep in mind that the plugs stand off the face of the ECM about $1\frac{1}{2}$ inches, so make sure there is room for the plugs as well as the ECM in your mounting location.

The LT ECM is dependent on the year and version of the engine. There are 2014–2016 ECMs, 2017–2018 ECMs, and 2019-up ECMs. Make sure to write the year, make/model, and VIN on the back of the ECM so that you always know where to find the VIN if it is needed.

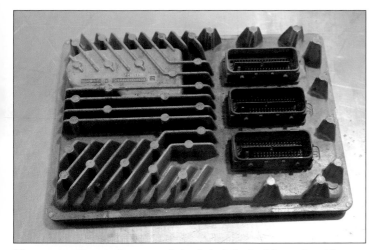

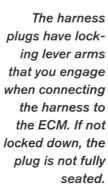

The harness plugs have locking lever arms that you engage when connecting the harness to the ECM. If not locked down, the plug is not fully seated.

There are three plugs, and each is labeled by color. You can't connect the wrong plug to the wrong color.

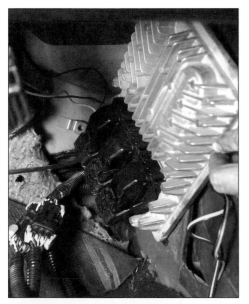

We mounted the ECM behind the kick panel on the passenger's side of the 1987 Camaro, just like the factory ECM. It is a tight fit though.

For the Buick GS, we mounted the ECM in a clean, discreet location: under the hood and behind the battery. We used a Dirty Dingo clamshell but had to fabricate our own mounts. We began with a sheet of aluminum and marked out an outline of the ECM mounting base.

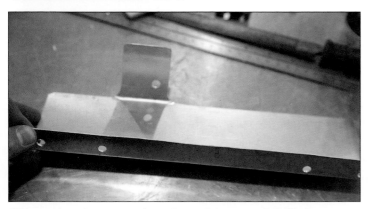

We cut the shape out, bent it to shape, and drilled some mounting holes. The lower flange will bolt to the Dirty Dingo clamshell; the upper small flange bolts to the forward core support, behind the headlights.

We used a piece of scrap sheet metal for the lower clamshell mount.

The two mounts were bolted to the clamshell and painted flat black. Then we added some studs for the clamshell front plate.

The front plate and ECM look good and fit well.

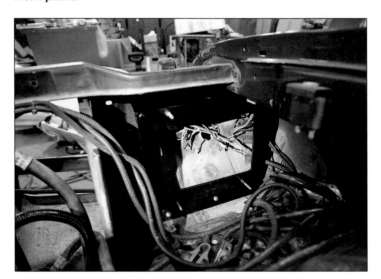

Our new ECM mount fits right in the core support. The battery will cover most of this.

With the ECM and cover installed, the bracket disappears. All that can be seen is the ECM.

Sensors

There are a few sensors that are required for the LT-series engine to operate correctly, and not all of these come with the engine. These include the fuel pressure sensor (part number 13579380), MAF sensor (part number 23262343), and oxygen sensors (two required, part number 12659516). How these sensors are installed and where they are located makes a difference in how the engine operates.

The MAF sensor must be installed inline with the air intake tube with some pretty strict guidelines. Unfortunately, this is just not always possible. GM specifications require an air inlet tube that is 4 inches in diameter by at least 6 inches long with the MAF mounted at least 10 inches away from the throttle body. This means that the section where the MAF sensor is mounted must be 4x6 with the MAF located in the center, which must also be at least 10 inches from the throttle body. So, the air inlet tube must be at least 16 inches long. It is not always possible to make this work, but it needs to be as close to this spec as possible. You can always have a longer air tube, but

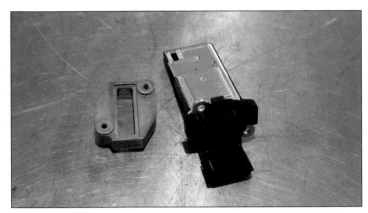

To control the engine, a mass air flow (MAF) sensor is needed. The one we used for our LT swap is part number 23262343. The mount that is needed can be purchased from ICT Billet (part number 551545), and it is aluminum. We decided to mount ours in the factory air intake on the 1987 Camaro.

We took the original air box and figured out a way to make it match up to the LT throttle body using a silicon adapter and a piece of stainless tubing. Is this perfect? No, it isn't, but it does work.

To get the tubing to fit inside the air box, we notched the end to match the male notch in the air box.

To make this a permanent change, the tubing is bonded to the air box with epoxy. We first sanded the interior of the air box with 80-grit sandpaper.

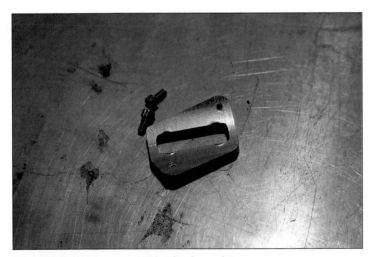

The MAF sensor mount is designed to go on a 4-inch-diameter tube, but we are mounting ours to a flat surface, so we need to sand the curve off.

This is critical: the top to the MAF is directional, so be 100-percent sure that it is in the right way. We marked the sensor and the mount to note the top and the airflow direction.

shorter is not a good idea.

Mounting the MAF requires a couple of parts. The MAF itself is long and rectangular with a large plastic cap for the plug. It does not come with a mount; that part is added to the air inlet tube. An aluminum mounting boss from ICT Billet (part number 551545 for $19.99) can be used to secure the sensor to the air inlet tube.

This is a weldable piece, but it is aluminum, so either the tube needs to be aluminum and it is TIG welded or epoxy is used to bond it to plastic or steel tubing. We used this mount on the 1987 Camaro featured in this book along with some rivets and epoxy to secure the MAF to the original air inlet manifold. Be sure to pay attention to the orientation, if it is wrong, the engine simply will not run well at all.

Throttle Pedal

To control the throttle, a throttle pedal or (in this case) an accelerator pedal position sensor is needed. Because all Gen V LT-series engines are drive-by-wire, an electronic pedal must be used. It can use a donor pedal from a 2014-and-up truck or Corvette, but for swaps, the simplest solution is the CTS pedal

We figured out the best spot for the MAF to receive the best flow and marked the air box.

Then we cut out the section. The two holes are for the rivets we will use to secure the mount.

Double-check all the fitments before riveting the mount in place. Before the rivets get fastened, we need to mix up the epoxy.

We mixed some JB Weld epoxy for plastic to ensure the epoxy would adhere to this type of polypropylene plastic. We put just a bead around the edges; we don't want the epoxy to cover the sensor hole.

Then we installed the mount to the airbox and compressed the rivets.

The MAF fits perfectly. We then added a bit of epoxy around the mount just to ensure it had a good seal. Finally, we painted the air box.

(part number 10379038). This is the unit that General Motors sells in all of its swap kits because it is considered as the best piece for a universal fit.

In every case, a bracket will need to be built to mount the pedal to the firewall. Sometimes this is simple; sometimes it is not. Whenever possible, try to use at least one of the original mounting points for the original pedal as a locating point for the bracket. This will ensure that the pedal is close to where it needs to be to feel comfortable when driving.

Tuning

Programming the stock ECM can be done several ways. There are home-based programs available, such as HP Tuners and EFILive, which are designed to work with the Gen V LT-series E92 ECMs. Most of the aftermarket tuning modules and software are locked, meaning they lock themselves to a particular ECM on the first use. To unlock them, you must purchase credits or VINs for additional vehicles. Some products include multiple credits or passes to tune multiple vehicles. Typically, when purchasing tuning software, you have unlimited tuning capability or tunes on the same vehicle.

Do-it-yourself or home-based tuning packages are very efficient and handy to have. They allow you to tune the engine for a multitude of parameters, help identify trouble codes, and add some personalization to the engine without getting greasy.

There are some things that home-based tuning software can't do. Most software packages can't make significant changes to the programming of an ECM, so you can tune the engine, but you cannot change the ECM from drive-by-wire to drive-by-cable or vice versa. The software also is not the best way to swap a computer from a different year to a different engine; this requires more significant programming beyond what the tuning software was designed for. These programs are great for what they do: tune the existing programming to better suit your application and needs.

HP Tuners

This comprehensive tuning software allows you to tune and adjust every aspect of the stock ECM. Available in a base or pro form, the HP VCM Suite adjusts Gen V ECMs. This system works on the credit system and allows tuning of up to four specific vehicles, unlimited tunes for the same year and model vehicle (any 2015 Camaro SS for example), and unlimited tunes to Gen V vehicles.

Whichever option is chosen costs credits. A single-vehicle license costs 2 credits, a year-and-model license costs 6 credits (most LT-series vehicles are not available in unlimited), and unlimited LT tunes cost 70 credits. The basic software purchase includes 8 credits, and additional credits are always available, adding to the flexibility of the software.

Throttle Pedal Mount Fabrication

Unlike a standard cable-operated throttle pedal, the throttle pedal position sensor is a large block mechanism that is mounted off one side of the main unit with three bolts. The pedal can be purchased new or used; some come with the factory steel bracket, and some do not. This bracket is not of much use other than to simplify the fabrication process. We used a pedal we sourced online. It came with the steel bracket, which we used as the base to weld up the rest of the mount.

Take a lot of measurements and make notes on where the pedal sits before removing the original pedal. We used the original gas pedal mount to make a pattern. This is the base for the new mount we will fabricate. We simply traced the base pad of the stock pedal mount onto a piece of poster board. We also marked the locations of the holes on the board. This was transferred to a piece of 1/8-inch steel, cut, and drilled.

The factory mounting bracket has tabs and extra material that gets in the way, so we trimmed away the mounting tabs (to the original vehicle), leaving only the main bracket. The angle of the pedal greatly depends on the shape of the firewall and the other obstacles in the way. Be sure to take careful measurements to get the new pedal to sit how you want it.

In the case of the 1987 Camaro we are working on here, the pedal mount sat at an angle of 32 degrees perpendicular to the motion of the actual gas pedal. If we just built the pedal straight off the firewall, the pedal would run at a severe angle, making it unusable. We factored in the angle and offset the bracket to compensate.

The depth of the pedal off the firewall depends greatly on the make and model of the vehicle, as well as where you prefer the pedal to sit. These pedals only have a couple of inches of swing, so they don't move very far compared to a stock pedal. You may be tempted

The CTS pedal is the pedal that Chevrolet Performance suggests for all LT swaps. We ordered ours from a salvage yard. It comes with a simple mount that we can hack up.

The stock Camaro pedal assembly is very simple, but adapting the new pedal will be tricky. We have to deal with a compound angle for the firewall, and the pedal itself operates on a separate plane.

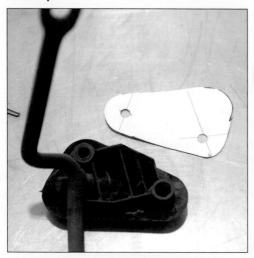

1 *First, we made a pattern of the original base that bolts to the firewall.*

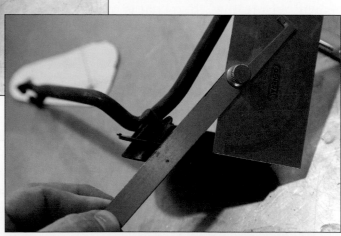

2 *Then, we measured the angle of the pad to find the correct plane for the pedal to move along. You don't want a pedal that swings anything other than straight forward.*

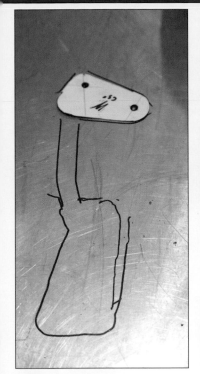

3 We used the bench to trace the pedal and laid out the first angle, which is to place the pedal vertically to the floor.

7 The bracket mount was boxed in to create a secure and sturdy mount for the new pedal.

4 Next, we cut out a piece of 1/4-inch plate steel and drilled it to match the original base.

8 With the bracket made, we wanted to adapt the original pedal pad to the new throttle pedal assembly. We removed the CTS pedal pad and set it aside.

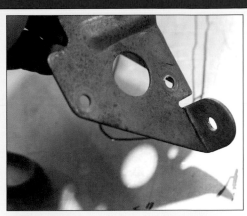

5 Then, we cut off a mounting tab that was in the way on the pedal mount.

6 We cut a small strap of metal and welded it to the base and the mount frame. The angles now match what we had previously measured.

to put the pedal as close to the floor as possible, but that will make the driving experience frustrating. You want the throttle pedal about an inch or so below the brake pedal, anything more is too much. You are looking for an easy transition from the throttle to the brake pedal.

With the depth of the pedal decided, we cut some 1/8-inch flat stock steel to length and tacked the new mount together. Ensure that none of the support stands obscure any of the mounting points for the pedal or the firewall mounts. Just do a couple of tack welds on each piece to test out the fit of the pedal.

The last step of this project actually isn't about the

Throttle Pedal Mount Fabrication *continued*

mount but the pedal itself. The pedal comes with the same basic injection-molded pedal pad that all new GM cars use. To bring the original stealth feel to the Camaro, we wanted to reuse the original 1987 pedal pad. We removed the pads from both throttle arms and tried to install the old one using the spring from the old pedal. It was close, but it didn't quite fit.

In the end, we had to shave about 1/8 inch off the top of the new throttle arm. Then, we added some extra material to the bottom of the arm and ground this all down smooth. The

purpose of this was to get the right angle on the pedal. The Camaro pad has a wider slot for the pedal arm, so we had to make a spacer using some number-10 screw washers.

You could easily cut a small piece of tubing to make a solid spacer, which is what we will do before we finish the build of the Camaro. The stock GM CTS pad does not pivot on the pedal arm, but with our new modification, the 1987 Camaro pad now looks and feels right. Plus, it pivots on the arm just like it did from the factory. You would never know it was not the stock pedal, even when driving the car. ■

9 We figured out that we needed to add some material to the base of the new pedal stub, so we stacked some welds on top of each other and then ground it all smooth at the angle we needed. This was to keep the pedal from flopping down.

10 Using the original pedal pin (from the Camaro), we mounted the pedal to the shaft. You can see that there is a large space in the center, that must be fixed.

11 With the spring and some small number-10 screw washers, the old pedal pad fits the new throttle assembly, and the original pedal look is retained.

12 Now, the spring-loaded pedal feels just like the original. Is this a step too far? Maybe yes, but maybe it is just far enough.

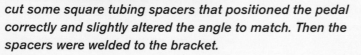

13 For the 1971 Buick GS, things were much different. The firewall is much closer to being perpendicular with the floor, and it is closer to the actual footwell where your feet rest. So, we simply cut some square tubing spacers that positioned the pedal correctly and slightly altered the angle to match. Then the spacers were welded to the bracket.

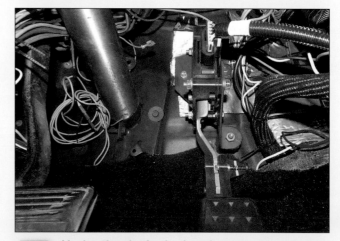

14 Under the dash, the bracket was mounted to the firewall. We may swap pedals on this one at some point as well.

This home-based software costs a good deal more than a handheld tuner or shipping your computer off, but it has much more flexibility and will likely be worth the extra cost in the end. However, the HP Tuners software cannot load programming from different model years to any given computer. Therefore, you need to start with the right year computer for your engine application.

Considered by many to be the most comprehensive tuning software publicly available, the HP Tuners software is an excellent choice. The HP Tuners Suite comes with the handheld scanner required to upload the tunes to the vehicle, but the handheld does not come preloaded with tunes. The software provides access to hundreds of tunes, and you can build your own. The product bought today will not be the same in two months, so the HP Tuners software is easy to upgrade to the most up-to-date information and calibrations.

EFILive

EFILive offers several versions of its FlashScan V2 products. FlashScan V2 for GM tuning is the product needed to work on GM Gen V ECMs. The tuning version allows you to build your own ECM tunes, controlling all of the aspects of the ECM programming. The FlashScan software works on a VIN licensing system, which requires each vehicle to be licensed.

The main kit includes two VIN licenses, and additional licenses can be purchased. The EFILive software is capable of tuning both Gen III/IV computers. The FlashScan software features more than 600 engine calibrations for the in-depth ECM tuning, as well as fixes for issues with the stock computers, such as VATS, MAF, and TCS fault codes.

The software is constantly upgrading and evolving. This ensures that the product you purchase will not only serve its current purpose but also be up to date as the technology changes.

Handheld Tuners

These are an alternative to the home-based software programs and offer more portability for tuning the engine. Handheld engine tuners allow tuning of the computer on the fly and running diagnostics whenever the need arises, regardless of location. Higher-end handheld tuners are typically comparable in price to the entry-level home-based software programmers.

Handhelds usually control the basic functions and parameters needed to tune the engines, but only the basics, meaning the computer-based software programs have many more variables and options. The biggest advantage of the handheld tuners is their ease of use. The simple user interface makes tuning easier than the home-based software. There are many handheld tuners available; the market is flooded with quality viable options.

Don't fret about tuning your ECM yourself. A swapped vehicle can be taken to a dyno shop, where the ECM can be tuned. This is a huge benefit because performance can be upgraded at the same time.

If you are just looking for an unlocked ECM, there are plenty of options there too. The main thing needed in an ECM is that the vehicle anti-theft system (VATS) is removed. This is the factory security system, and it must be turned off for the vehicle to start. This can be done with HP Tuners and EFILive, but most handhelds cannot perform this operation.

Most aftermarket harness makers and some specialty retailers offer ECM programming services. Options are to send in an ECM or purchase a new ECM that has already been reprogrammed with the VATS deleted, adjusted for your gearing, emissions controls deleted (rear cats, purge canister delete, fuel tank EVAP system delete) and already tuned out of the system so that you can just connect and fire up the engine. It is a good idea to get the engine dyno tuned after the installation so that your engine will operate correctly.

DOD Delete

Many swappers who have completed a Gen V swap report that the factory displacement-on-demand (DOD) system should be deactivated in the ECM. The common issue is that in a swap application, there are fewer chassis and drivetrain sensors, which make the DOD system not work very well. It also tends to be noisy.

This can simply be turned off on the ECM for stock engines. However, if you perform a cam swap, then the DOD system must not only be deactivated in the ECM but also the hardware in the

Painless Performance's Perfect Torc Transmission Controller Installation

Unlike the earlier *slushbox* automatic transmissions that soaked up massive amounts of power and left you with sloppy shifts and slower ETs, the modern electronically controlled transmission operates with much tighter parameters. Considering that 0–60 times for most high-performance vehicles are now spec'd with automatics instead of their manual-shift counterparts, the reality is that the automatic transmission is the better solution. That also means that an electronic control unit is needed.

For LT swaps, this creates an issue because most of them do not have the transmission controller built in. You need a separate controller. Yes, you read that correctly, the LT-series engine ECM does not have transmission controls built in—it is a separate controller. A laptop is required for tuning the controller.

Painless Performance has the solution, and it is called Perfect Torc. Designed to manage GM 4LXX transmissions from 1995-and-up (1995-and-up 4L80, 1996-and-up 4L60), this little unit packs a lot of features. Not only does it give tuning control of the transmission (including shift timing and firmness) but it also comes with push-buttons for paddle-shifting as well.

For the most basic installation, there are only four wires to connect: ignition (handles all power functions), two grounds, and the TPS sensor input. The rest of the wires are secondary controls, including button-shift, speedometer output, and a 5-volt TPS power feed for carbureted engines (not used with an LT). The other wires are all bundled together in two plugs that plug directly into the ports on the transmission.

The installation is quite simple. Under the vehicle, route the transmission harness to the passenger's side of the transmission case. Plug the large round plug into the port on the shelf of the case. There is a second two-wire plug that connects to the rear speed sensor port. Make sure that the wires are routed away from the exhaust and driveshaft. The main harness is long enough to reach the firewall and into the vehicle for most cars and trucks.

You could also run the harness through the floor—just be sure to use a grommet to protect the wires from chaffing. This harness has three separate molex connectors. The controller itself should be mounted inside the vehicle. The glove box or center console are great locations because the controller has a diagnostic and tuning function for on-the-fly tests and tuning.

The controller harness has eight wires in one molex connector. These are the power, grounds, sensors, and shift button wires. As previously stated, only four wires are required for operation. The power wire terminates to a 12-volt source when the ignition is on. This must have power only when the ignition switch is in the run/start position. The ignition switch is the best source for this wire.

The grounds must be terminated as close to the ECM ground as possible. Most LT engine harnesses have multiple ground points, so locate the closest set and run the ground wires to that point. On our vehicle, the grounds were made to the driver-side valve cover with the ECM grounds; this was directly in front of the large factory grommet in the center of the firewall.

The TPS signal wire can be tricky. If there is no pinout for the ECM, you will need to trace this wire. Our install is using an LT1 ECM. According to the GM ECM wiring diagram in

For swaps that do not have a transmission controller in the ECM, such as LT swaps, you will need one if you plan on running an automatic transmission. This is the Perfect Torc from Painless Performance, and it is a very simple install.

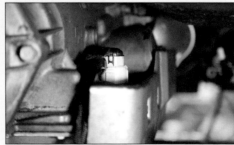

There are two harnesses: the transmission harness and the module harness. The transmission harness routes under the vehicle to the transmission. The main bulkhead plugs into the port on the top of the pan on the passenger's side.

The tailshaft VSS sensor plugs into the harness on the passenger's side of the tailshaft. Make sure that the wires are routed away from the exhaust and any moving parts. The transmission harness was run into the vehicle through a factory hole in the firewall.

Inside the car, there are several wires to connect. This bundle includes the optional wires for bump shifting and speedometer output. We can use these later, but they are not necessary for the initial install.

Because the transmission controller keeps the unit working, we opted to solder the power connection. It can be crimped, but if the crimp fails, then the transmission can fail too.

The Perfect Torc needs a TPS signal to function properly. We used the GM wiring guide to locate the TPS signal wire inside the main engine harness.

To verify we had the right wire, we checked the TPS plug and found the same color-coded wire.

Using a pair of auto-strippers, a 3/4-inch section of the wire was stripped. We used a razor blade to remove the stripped insulation completely.

Next, we put a test wire into the TPS signal port and touched one lead for a multimeter set on continuity.

We then touched the other lead to the stripped wire. This is just a precaution to make 100-percent sure we have the right wire.

chapter 6, port 70 on the C2 or J2 plug is the TPS signal. LT engines typically have two TPS signals (redundant systems) that need to be connected to the actual signal side. Check this with a voltmeter; the wire that changes voltage with moving the throttle open and closed is the correct signal wire.

The speedometer output can be calibrated to operate any electronic speedometer. The calibrations are all managed in the laptop software that is included with the system. Mechanical speedometers are not supported by GM 4LXX transmissions. If you must retain the mechanical speedometer, then an electric-to-mechanical conversion unit is required. The vehicle speed calibration can be done with a GPS unit, a pace vehicle, or through an adjustable speedometer.

If you want the ability to paddle-shift your vehicle, there are two wires on the harness for that purpose. The kit comes with several push buttons for dash or console mounting, or a steering wheel adapter with paddles can be purchased. Three buttons are required: one wired to each shift wire (turns on/off manual mode), and then each shift wire is connected to a separate switch for shifting. The other side of the up/down buttons goes to ground. Each press of the buttons shifts the transmission up or down one gear.

The last wire on the controller harness is for table selection. The Perfect Torc unit can manage two shift tables, allowing the

Painless Performance's Perfect Torc Transmission Controller *continued*

user to set up a performance mode and a race mode, or street and off-road shifting tables. This uses a toggle switch (included). When this wire is grounded, table 2 is used; when it is open, table 1 is used. While it is not required, this is very useful for dual-purpose vehicles.

Once the wiring is completed, simply plug in a laptop and run the software. The program has several base tunes to choose from and then tweak from there. Load the tables and tune, and then enjoy a freshly tuned transmission. ∎

The trick to soldering a second wire into the middle of another is to split the continuous wire and slide the new wire between.

Then, wrap the wire around the continuous wire and solder. The splice is completed by wrapping the joint with electrical tape.

We reinstalled the wire loom and taped up the joint where the new TPS signal wire exited the loom. We used Painless Performance's Power Braid to conceal the signal wire.

The Perfect Torc system uses two ground wires, which must be grounded close to the ECM ground. The one pair of ECM grounds (there are two pairs total for the LT1 ECM) and both Perfect Torc grounds were secured to the rearmost driver-side valve cover bolt.

engine must be removed as well. This is due to the fact that most aftermarket camshafts are not designed to work with the DOD system.

Transmission Controllers

The 2014-and-up Gen V LT-series ECMs do not have transmission controls built in. If using an electronically controlled transmission, a separate transmission controller is needed. Chevrolet Performance offers the "Connect & Cruise TC-2" transmission control unit (TCU) system for 4LXX

series transmissions. There is a kit for 4L6X/7X units and one for 4L8X transmissions. The TCU plugs into the ECM so that the two controllers can talk to each other. These transmission controllers also work with E-Rod and other Chevrolet LT ECMs.

The factory TCMs are great, but there are aftermarket versions available as well. There are several aftermarket transmission controllers, such as the TCI TCU and Painless Performance's Perfect Torc, which provides greater tuning capability of the GM electronically controlled transmis-

sion than the factory controller. These units are compatible with most GM electronically controlled automatic transmissions. It provides load, gear, RPM, and speed-based programming with the click of a mouse.

Transmission controllers come with the tuning software and wiring harnesses. The software allows you to adjust many parameters, even allowing paddle and push-button shifting configurations and manifold-pressure-based shift firmness to tune the transmission far beyond the capabilities of the stock TCM or ECM.

FUEL SYSTEM

The Gen V LT-series engines are a little bit different from the LS series, particularly in the fuel-system department. All LT-series engines are direct injection (DI), meaning the fuel is pressurized between 2,000 and 2,900 psi (2,175 for the LT1 and 2,900 for the LT4) and injected directly into the combustion chamber, much like a diesel engine. DI engines have much greater fuel economy and more power poten-

tial because the ECM has far better control of how much is being burned. This is due to the three types of combustion modes in the ECM: ultra-lean, stoichiometric, and full power.

Ultra-lean mode is used during cruising, when there is no acceleration and light loads. The engine may see air/fuel ratio (AFR) readings as high as 65:1. This is possible due to the swirl chamber design in the combustion cham-

ber. The highly pressurized gasoline is injected directly at the spark plug, keeping the heat away from the cylinder walls. The fuel injection is also performed later in the compression cycle. This burn is much cooler than in a conventional engine, which is how they get away with such high AFR figures. This is the hyper-economy phase of the engine.

Stoichiometric burn is what we all know to be the typical burn mode. An AFR of 14.7:1 is

The easiest way to upgrade the fueling system to match the needs of the LT-series engine is to replace it with a new tank that has been retrofitted with an electric fuel pump. The pump needs to be capable of flowing 75 gallons per hour (gph) at 45 psi and be PWM controlled. This Phantom fuel system from Aeromotive is for a first-gen F-Body, and it works perfectly for an LT out of the box.

Another option is to install a new pump using a retrofit pump kit like this one from Holley. The same requirements apply. It has to be PWM controllable and capable of flowing 75 gph at 45 psi. (Photo Courtesy Holley Performance)

the target for the ECM. The fuel is injected during the intake stroke, generating the type of air/fuel mix that is experienced in a conventional engine.

Full-power burn is just like the stoichiometric mode, only more fuel is added to keep detonation at bay. The LT1 has 11.5:1 compression, so this is a critical function for hard acceleration. In the LT4 supercharged version, compression is lower at 10:1, to reduce cylinder pressures with the 9.6 psi of boost.

To achieve these burn modes, there are two fuel pumps: an electric supply pump in the tank and a mechanical pressure pump underneath the intake. The mechanical pressure pump runs off a tri-lobe wing on the camshaft. Aftermarket upgrades to the fuel pump lobe can be made through the camshaft, which can increase fuel flow as much as 74 percent. The electric supply pump in the gas tank is different from a standard electric pump as well because it is PWM controlled.

Instead of a basic fuel pump and regulator, the factory supply pump is controlled by the ECM through a fuel pump module. This controls the base fuel pressure as it reaches the mechanical DI fuel pump. A special pressure sensor in the fuel line monitors the pressure of the fuel to maintain adequate pressure for the current engine demands.

The primary spec that General Motors states for LT-series engines is 72 psi at 45 gallons per hour (gph). However, this is slightly misleading because this number is for high-demand peaks. Under part throttle, the LT fuel system runs as low as 54 psi for cruising. Rather than use a regulator, the pressure is managed through PWM control of the pump.

Essentially, the ECM switches the voltage and current sent to the pump on and off at a very fast rate to control the speed of the pump, ensuring full pressure at all times with no delays. This complicates the fuel system for the LT-series engine swap. Not just any old fuel pump can be used. DI fuel pumps have to be PWM capable, and not all electric fuel pumps are.

Another factor is that the LT fuel system is returnless; this was done to keep the fuel temperature down. Because hot fuel does not cycle through the pump to the engine and back to the tank, the fuel temp remains even. Returnless fuel pumps are rarely suitable for EFI use without PWM control, and the ones that are available cannot support the type of pressure and flow needed for an LT. The pump requirements are 72 psi at 45 gph and an 84-psi pressure relief to be compatible with the Chevrolet Performance control system, which is a pretty high burden for a street-driven electric fuel pump.

It is possible to run an LT-series engine without the PWM control. There are a couple of reasons that swappers may choose this route, and the main one is to not deal with an in-tank pump and the extra wiring. In reality, the static system is just as complicated for most applications because a return line has to be added and the pump has to be wired to function.

Plumbing the fuel system for static pressure is done with a standard regulator and return line system set to 72 psi. This process is used for engine dyno testing, which is often done at 60 psi, according to Lingenfelter Performance. There are a few issues with this in a street car. First, 60 psi is not 72 psi, so the DI pump will not be functioning at full capacity, meaning that it will not be full of fuel under full load. While this works on the dyno, real-world street applications could run into some drivability issues, as has been reported by some LT swappers.

If the regulator is set to maintain 72 psi, the system should work as intended, but the return line can cause some issues with fuel temperature. Because a return line sends fuel back to the tank, the result is an incremental rise in fuel temperatures. While vapor lock should not be an issue, hot fuel at 2,000 psi becomes more volatile, which can cause some issues.

Many retrofit systems drop the hot returning fuel directly in the flow path of the fuel pickup. This means that most of the returned fuel gets sent right back through the pump and to the engine, where it gets returned again, this time even hotter. If you choose this route, the return line needs to be located as far from the pickup point of the fuel pump as possible.

A PWM system does not need a return line at all because the fuel control module slows the pump down when less fuel is needed. It can instantly kick up to max pressure in milliseconds. This ensures the mechanical pump always has the supply it needs.

Fuel Pumps

When it comes to fuel pumps, there are two categories: in-tank or external. The in-tank pumps are more complicated to install, but they tend to last longer, hold higher pressure, and run quieter. Retrofitting an in-tank pump costs more, but the cost is worth it in reliability.

There are two ways to add an in-tank pump to a vehicle: buy a tank with a preinstalled EFI pump (either custom or OEM from a different vehicle) or retrofit the tank. Both options can be pricey, but there are ways to keep it on the cheap as well.

An in-tank pump that can supply the 72 psi at 45 gph fuel-pressure requirement is required. The Aeromotive Phantom 340 in-tank retrofit pump has the capabilities to support the LT fuel system and is quite easy to install.

Another option is to use a factory GM fuel-pump module from one of the LT-powered vehicles from 2014-and-up that were used in 1/2-ton trucks, Corvettes, and Camaros. The Camaro fuel module is fairly tall, so it does not fit in most car tanks, but the 2014-and-up GM truck fuel modules fit quite well. Installing one of these pumps requires some fabrication on the fuel tank. The nice thing is that if you use a factory module, the wiring for the pump is basic plug and play.

New tanks can cost $600 to $1,000. Companies such as Rock Valley Antique Auto Parts build custom stainless steel gas tanks for street rods, hot rods, muscle cars, and trucks. If your car is not

For the 1987 Camaro, we went with option C, which is to install a new pump in place of the original in-tank pump. If the vehicle had EFI from the factory, such as almost any 1987-and-newer GM vehicles (and a few older models), then it already has a fuel pump in the tank. This upgrade begins by knocking out the retainer ring for the pump assembly.

We went with a Deatsch-Werks DW300 pump kit. This pump can flow the required 75 gph at 45 psi and is capable of using a PWM signal. It is also the same size as the original in-tank pump.

The new pump is shown next to the old. There are a few things that we won't be using, including the wiring, connectors, and the return line.

First, we removed the old pump and disconnected the wires. The black connecter is also removed.

On the underside of the pump assembly is a little vapor line. We removed the cover and the keeper. This will be the wire run.

Up top, the line that went to the vapor filter was cut off. There is still a vapor line attached, but we needed one of these for the port.

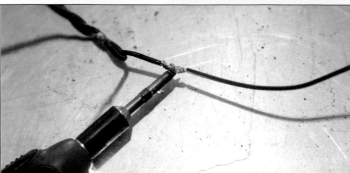

This connection must be soldered, **do not** crimp this. This is a PWM signal wire, and it must be a proper connection.

The pump comes with a plug pigtail. We need to lengthen this and seal it from the gas.

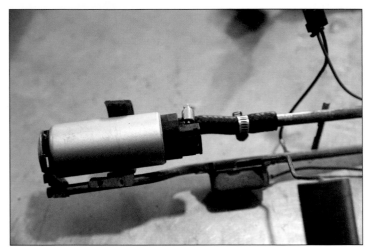

The solder joints were covered with shrink tubing and then sealed with fuel-resistant silicone for an extra layer of protection.

We mounted the new pump to the assembly with a section of fuel line and a worm clamp. The pump will be secured with a couple of hose clamps to the no-longer-used return line, which is also the support stem.

We routed the pump wires out of the top of the assembly and coated the entire area with some more fuel-resistant silicone.

Don't forget to put the filter sock onto the new pump. We set the pump in the stock location, so there is no concern about the filter not reaching the bottom of the tank.

The unused lines get capped with silicone caps.

on the list, Rock Valley Antique Auto can build a custom tank. These new tanks are very slick, well built, and easy to install, but they come with a high price tag.

Each Rock Valley EFI fuel tank features a dropped sump in the top of the tank. This allows for adequate floor clearance. Each EFI tank comes with a new high-volume, high-pressure fuel pump to feed the engine. For the most popular vehicles, Aeromotive and Holley offer replacement tanks that are closer to the $600 price range, and they use nice reproduction tanks with their own proprietary pump systems.

Ambitious builders can choose to modify the stock tank themselves. Note: The following procedure requires welding on a gas tank. Serious injury or death can occur if the utmost care and preparations are not followed. There are a several ways to alter a stock tank to take an EFI fuel pump.

The bargain-basement method is to take the stock sending unit assembly out of the tank, cut off a short section of feed line (about 1½ inches, depending on the depth of the tank and the length of the pump), and fit the pump to

the stub using fuel line and hose clamps. The important thing here is to make sure the pump is mounted about 1/4 inch off the floor of the tank with a filter sock resting on the bottom of the tank. This keeps the impurities out of the pump while getting the most fuel out of the tank.

The pump needs to be supported, so a piece of steel rod can be welded to the underside of the assembly plate. The pump is then clamped to the rod, so it remains stationary. This method won't work for all tanks, especially shallow tanks, and it may be hampered by the diameter of the stock assembly. Additionally, the fuel-level sending unit may be in the way, depending on the application. This method keeps the stock feed lines in place and eliminates any floor pan clearance issues. One more drawback is the complete lack of a fuel sump feature, which traps fuel around the pump inlet, ensuring it does not run dry. Running an in-tank pump dry is very destructive; they don't last long when run dry.

The next option involves welding on the tank. This is extremely dangerous and should

not be attempted in haste or by novices. All of the old fuel must be removed and the tank thoroughly rinsed, drained, and rinsed again until there is absolutely no possibility of any remaining fuel vapor. If you smell a hint of gas, do it again. Some people even suggest filling the tank with water or inert gas, such as argon, while the welding is being performed. In any event, when in doubt, seek the help of a professional. Most fuel tank builders offer their services for retrofitting tanks, so employ their services if you can.

Installing a custom in-tank pump in the top of the tank often requires a recessed panel on the top. The fittings clear the floor pan and provide a flat surface to mount the new assembly. If you place the top sump to the side of the original, the stock sending unit can be used, simplifying the process. This requires a boxed section be built and welded to the top of the tank. Then a fuel pump assembly unit is built with both wiring and inlet and outlet fittings. This piece should have a bar or rod on the inside portion of the tank for the pump to mount to.

Using 90-degree hose barb fittings is usually the easiest way to get the fuel in and out of the assembly. These fittings must be sealed tight so they don't leak. The entire assembly bolts to the sump. In addition, by installing baffles to the inside of the tank, fuel will surround the pump at all times. These should be added before the top sump is installed.

The other option is to weld a sump into the bottom of the tank. This sump would be placed directly below the pump, but the pump would be installed in the lowered section, with the filter sock on the floor of the sump.

Tanks Inc. offers an upper tank mount, complete with a fuel pump and a baffle. This option reduces some of the legwork in building this piece and ensures the pump will be covered with fuel at all times.

There is one more option that allows you to install a truly high-performance fuel pump into a stock fuel tank without welding. Aeromotive offers two retrofit in-tank pump kits: the Phantom and the A1000 Stealth systems.

For basic street performance use, the Phantom 340 kit is suitable. This kit allows you to simply cut a hole in the top of the tank, drop in the pump, bolt it down, hook up the lines, and you are done. The kit comes with the seals, hardware, and a drill jig to ensure the holes are in the right place. The 340 supports up to 700-hp supercharged EFI engines or 1,000-hp supercharged carbureted systems. The Phantom system can fit in just about any tank, so this is a really good option that takes out the guesswork.

For more serious performance engines, the Stealth A1000 system feeds up to 1,300-hp EFI systems and 1,500 for carbureted engines. Installing these systems takes slightly more effort than the

Next, we need to make a flex line to connect the fuel line to the engine. We are using Earl's Performance Plumbing fittings and Super Stock hose, which is a push-on style. It can handle 250 psi, good enough for our LT. We also have an adapter that will push onto the stock LT fuel rail. As a side note, always use swept tube–style fittings like this for fuel; the block-style 90-degree fittings cause turbulence in the fluid flow.

A little assembly lube makes pressing the hose onto the end much easier.

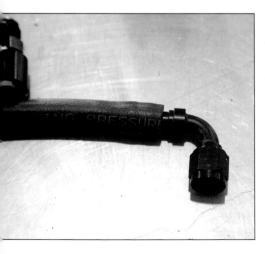

Then, the hose is just pushed onto the fitting until it won't go anymore. These may not seem like high-pressure hoses, but they work very well. Be sure to use the Super Stock hose with Super Stock fittings for this to function correctly.

One end of the hose was threaded onto the adapter on the hard line. AN fittings do not require much torque, just hand tight and then a slight tightening is enough. Too much torque will damage the seal.

Phantom kit, but not much.

External or inline pumps offer a simpler installation and are usually cheaper. Inline pumps are much easier to change, making roadside swaps bearable. The main gripe over inline pumps is the noise. Drivers may hear the whir of the electric pump over the engine with a stock-style quiet exhaust. For most builders, the added noise is merely an inconvenience. However, for a show car, a noisy pump might be considered a serious drawback.

The real drawback for an inline pump is that the fuel line is only pressurized after the pump, so the tank to the pump is gravity fed. Anyone who has dealt with a modern high-performance external fuel pump can tell you that life is really difficult when you lose the siphon in the tank. Simply having the pump in the tank maintains a constant supply of fuel to prevent those hard-cornering and acceleration woes that come with a stock tank and an inline electric fuel pump.

Inline fuel pumps also require a more substantial return line system. This is because of the long distance between the regulator and the fuel tank. Inline pumps are also subject to failure through heat. The only thing that cools the pump is the gas flowing through it.

Not all inline pumps are created equal. External pumps come in all different shapes and sizes with the majority of the market consisting of low-pressure units designed for carburetors. These pumps deliver 6 to 14 psi, and they are not close to the 72 psi required to operate an LT engine. Inline pumps are more susceptible to overheating, and most are not capable of PWM returnless operation. So, running this in a return-style system with an LT engine will result in more complications due to overheated fuel.

Installing an inline pump is pretty simple, but there are a few caveats. The first is to always install a prefilter before the pump, so the pump does not get clogged and ruined. A prefilter is a screen-style filter that traps the big stuff. A micron filter should be placed after the pump to catch small contaminants. Do not install a micron post filter in front of the pump (between the tank and the pump) because it will impede the gravity feed, and there will not be enough force to push the fuel through the micron filter.

Make sure the prefilter is large enough to free-flow the fuel. A small prefilter will restrict the flow to the pump, causing cavitation that will burn up the pump. A stock-type metal canister prefilter works great, but they are not pretty. Most aftermarket fuel pump makers have large-capacity prefilters if you want a good-looking filter. Most of the aftermarket prefilters are rebuildable as well.

We don't want to chop off the quick-release fitting on the fuel rail, so we are using a push-on adapter. This will connect to the fuel rail and leave a -6 AN connection on the other side.

All done, the fuel system is now complete.

Cavitation Explained

Cavitation is a natural process that occurs when vapor bubbles are induced in liquid under pressure. All electric fuel pumps are susceptible to this force. Its most common cause is installer error where an inadequate fuel supply increases the suction on the inlet side. Cavitation is literally boiling the fuel through pressure. Vapor bubbles form and then split, causing a micro-explosion. This is extremely damaging, and even a few minutes of cavitation can ruin a fuel pump.

Another cause is overheated fuel. Not running a return line (deadhead style) or attempting to plumb the return line into the feed line (not into the tank) will cause hot fuel to cycle back through the pump, heating it up more. The hotter the fuel, the easier it is to cavitate or even vapor lock. Yes, EFI systems can vapor lock too. This is why a proper return system is so important for EFI unless you are running a PWM-controlled pump. If you use the OEM-style PWM pump and deadhead-style lines, the pump only pumps the fuel needed on demand, so there is no need for a return line. ■

The most important aspect of any electric fuel pump is the wiring. In addition, it is difficult to get solid grounds because paint, rust, and scale inhibit the ground. Always be sure to remove the paint and anything else from the ground location, so there is clean metal.

Electricity requires equal grounding and positive current flow. A bad ground is just as bad as a faulty positive feed. Electric fuel pumps require a lot of current. Running a relay circuit from the pump trigger lead, along with 12-gauge positive and negative wires to the fuel pump, provides ample capacity. This ensures that the pump gets the required amperage without overheating the wires. Do not run a 16- to 24-gauge primary wire to a fuel pump because it will cause a fire. All fuel pumps require at least 12-gauge power wire, and the larger pumps need 10-gauge wire. This includes both power and ground wires.

Fuel Control Module Pressure Sensor

Regardless of how the fuel system is plumbed, the factory fuel module is required for an LT-series engine to operate correctly. The ECM must know the inline fuel pressure coming into the DI pump—without it, the engine will run erratically. To send this information to the ECM, the fuel module is needed. Because of this, running the factory-style PWM control is a matter of three wires. It is easier to use the PWM system than a regulator with return line system.

The factory fuel module is a learning unit; it has the ability to slightly adjust the fuel trims to match the pump it is controlling. This technology is still fairly new, so the aftermarket has not quite caught up yet. It is possible to tune the fuel module to match the pump being used in the system.

A special pressure sensor in the fuel line monitors the pressure of the fuel, which is maintained at 72 psi. Rather than use a regulator, the pressure is managed through PWM control of the pump. Essentially, the ECM uses PWM to control the speed of the pump, ensuring full pressure at all times with no delays. This complicates the fuel system for the LT-series engine swap. Don't use just any old pump. The pump must be able to be PWM controlled. Before purchasing a pump, make sure it is PWM capable.

Installing the PWM controller is relatively simple, but the fuel pressure sensor is a bit tricky. First, an inline adapter with a pressure sensor port positioned at 90 degrees or 5 to 85 degrees to the flow of fuel is needed, according to the GM manual for the fuel controller. This means that the sensor itself must be mounted either 90 degrees vertically or between 5 and 85 degrees from the vertical position. Essentially, the sensor should not be horizontal or below the flow of fuel.

As long as the sensor is angled upward (with the terminal sit-

If you want your LT engine to run like it is supposed to, use the fuel control module (FCM). This small computer controls the fuel pressure your engine receives based on demand.

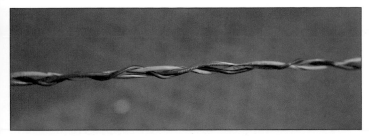

There are three wires that run from the FCM to the fuel pump, and they must be twisted with a shield ground wire. We braided our wires, but the minimum is 27 twists per 8 foot. The cleaner and more consistent the twist, the better the signal wires are protected. A loose braid is best because a really tight twist or braid can stress the wire itself.

The FCM must be mounted between the engine and the fuel tank; the Chevrolet Performance wiring harness has a very short pigtail for the FCM to fuel pump run, so it may need to be extended. On the Buick GS, we mounted the FCM under the car to the transmission crossmember.

There are several versions of the fuel pressure sensor that can be used with LT-series engines. Upper left is the Corvette and truck sensor; bottom right is the sensor used with Camaro LT engines. They are the same internally, but the Camaro sensor has hose barbs, while the other requires a 10-mm male threaded adapter.

Depending on the fuel system plumbing, the Camaro sensor may work better because the truck/Corvette sensor can become a bit clumsy if you can't find a direct fitting adapter. If you are using AN fittings, the Corvette sensor works better.

ting above the fuel port) at least 5 degrees, fuel cannot pool in the sensor. This is fairly easy because there are plenty of fuel sensor adapters out there. The problem is that most of the adapters are for 1/8-inch NPT fittings and not the 10-mm threads required for the GM sensor.

Finding a 1/8-inch NPT male to 10-mm male adapter is difficult. It is easier to find a -6 AN male to 10-mm adapter. To use this, you need an aluminum fuel log or Y-block fuel splitter and a -6 to 10-mm male-male adapter.

This allows you to connect the sensor into the fuel system. We made one with a leftover piece of fuel rail from another project.

Wiring the PWM pump controller is a plug-and-play affair, but the pump wiring itself is not. There are three wires coming off the pump module: yellow with a black stripe, gray, and a smaller-gauge black wire. The yellow/black wire is the ground, the gray wire is the power side, and the small black wire is the shield. If you are using a GM pump with a shield pin, connect the small

black wire to that pin, but if you are using a pump without a shield pin, leave the wire unterminated and tape it to the other wires.

Because of the nature of PWM control, there is a very real potential for electromagnetic interference (EMI) from other electronics in the car. To eliminate this from interrupting the control signal, the two main power control wires are twisted with a third shielding wire. This wire is grounded to the chassis near the pump.

The Chevrolet Performance wiring harness only comes with

Stepping Up to Steel-Braided Fittings and Hoses

While regular high-pressure EFI-spec rubber hoses are adequate for factory installations, braided lines and hoses provide the best look, performance, and fit for a custom install. For decades, steel-braided lines have been the standard for muscle cars and trucks, and they have been cutting fingers and being the bane of existence for many street rodders. There are a few tips and tricks that can take the strain out of plumbing your ride and getting the best look possible.

Earl's Performance Plumbing, a division of Holley Performance, offers a few products that bring the braided look under the hood while reducing the strain of stainless steel. The Pro-Lite 350 hose is a high-quality rubber hose with a black nylon braid that protects the rubber from chaffing, flying debris, and cuts. The nylon is easy to cut, saves your fingertips, and adds a serious look to the plumbing. The dark braid look is all business, but it really cleans up the engine bay. The Pro-Lite 350 can be used with all the AN fittings and simple hose clamps, adding to its versatility.

Working with AN fittings requires some extra knowledge and planning of each system. Typically, the local parts store doesn't carry AN fittings, so unless you have quick access to a speed shop, ordering fittings is the only option. Have the system planned out beforehand, so you don't waste time with parts that aren't needed or end up waiting for the ones you don't have.

The aerospace industry developed these fittings, and AN represents Air Corps-Navy. Each AN size directly correlates with a specific OD of metal tubing. Each size is listed as "-X."

Each number inside the quotation marks indicates a 1/16-inch increase in size. Therefore, a -3 fitting would be 3/16 inch, -4 would be 1/4 inch, and so on.

If the car sees a lot of street time, running hard fuel lines is recommended. Even braided lines can collapse over time, and a properly routed steel line will last almost forever. Depending on the stock system, it might be possible to reuse the stock hard lines. A 3/8-inch feed line and 1/4-inch return lines (if using a return-style system) are the minimum requirements for a high-performance fuel system. These sizes are good to about 500 hp. If your LT engine produces more than 500 hp, you need 1/2-inch lines.

There are two ways to convert from hard lines to braided lines with AN fittings. One way is to install a hose clamp on a bare hose. The proper way uses a tube sleeve, a tube nut, and an AN fitting. There is a trick to it though, and if it is not done properly, the fitting will leak.

The automotive industry standard flare is 45 degrees. This is the angle of the inside lip of the flare, and it is what seals against the fitting. All commonly available flaring tools are 45-degree kits. AN fittings, however, require 37-degree flares. Those 8 little degrees can mean the difference between a proper seal and a nasty leak.

Finding a 37-degree flare tool might prove to be a little difficult because local parts stores don't carry them. We sourced ours from the Matco truck, and it had to be ordered. A basic double flare is all that is needed once you have the right tool for the job.

1 *Running AN lines means assembling the fittings to the lines through a specific process. Here is a set of -8 fittings, -8 hose from Earl's, some Royal Purple assembly lube, and some AN wrenches.*

2 *The main fitting was lubed to accommodate the hose and the threads on the collar.*

3 *Next, the collar is installed on the hose. The hose should be 1/16 inch from the threads on the inside of the collar. Tape is wrapped around the hose at flush with the base of the collar.*

Planning out the system requires a little work and can be a tedious task. Careful measuring is important because you don't want to order 25 feet of hose only to end up needing 26. Using a simple drawing of a car and marking out the route for the lines and hoses can greatly simplify the task.

Determining the placement of the fuel pump is key for fuel systems. For the transmission cooling lines, route the hoses through the frame whenever possible to reduce the amount of exposed line. If a sharp turn is needed, figure in an angled fitting, such as a 45 or 90 degree, instead of trying to bend the hose. A 3-inch radius is the maximum bend for rubber hose.

Choosing which components to use depends on your budget and the desired level of performance. Earl's Performance Plumbing offers several different types of fittings and hoses to suit each system's needs. Black Ano-Tuff hard-anodized fit-tings resist corrosion and wear better than the more common red and blue anodizing, and these fittings are perfect for Earl's Pro-lite nylon braided hose.

Pro-Lite 350 hose is best suited for the fuel, heater, and transmission hoses because of its light weight and flexibility. Stainless steel Teflon lines or nylon-braided high-pressure hoses are needed for power steering lines. Swivel-Seal hose ends add flexibility to the lines because these ends keep the hose from twisting when assembling the lines in the car, otherwise they can collapse.

Often, there will be things that don't fit or need additional fittings. If you order parts online or from a catalog, you will have to wait for them to arrive before planning your system. Discuss your plans with the tech advisors; they will help you get exactly what you need. Once all of the lines are fit and installed, test each system for leaks before setting out on a road trip. ■

4 *Then, the main fitting is clamped into a vise. We are using a set of Earl's aluminum jaws to protect the finish of the fitting.*

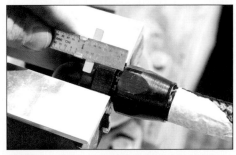

5 *Installing the hose to the fitting takes a little effort. Push the hose into the fitting and then spin the entire hose/collar assembly until it can't be threaded by hand anymore. Use the AN wrenches to seat the collar 1/8 inch from the end of the fitting.*

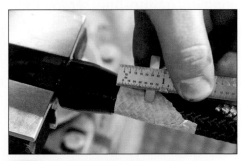

6 *The hose should not have pushed out from the collar more than 1/16 inch. If it did, reinstall it.*

7 *Most people just stop here, but the fitting needs to be pressure tested. Using a kit from Summit Racing, the ends were capped. One end has a Schrader valve for adding air.*

8 *The line was pressurized to the capacity of the system. In this case, about 65 psi. Check the hose at or above the pressure of the system.*

9 *While holding pressure, the assembly is placed in a jar of water. Test both fittings. Let the hose sit for 5, 10, and 20 minutes, checking to see if the hose loses pressure over time.*

a certain length, about 6 feet. To maintain the shielding, the wires must be twisted at a minimum of 27 twists per 8 foot of wire. The best way to ensure that the wires are correctly twisted and won't unravel is to braid the three wires together. It does not need to be a tight braid, rather a consistent loose braid, wrapping the wires around every 3 inches or so. Do not use crimp connectors for these wires. Instead, make sure to solder them well and use shrink tubing.

Return Lines

If you decide to go with a regulator/return-line setup, then you will need to run a new line. The minimum requirements for EFI fuel lines are 3/8-inch line for the feed and 5/16-inch line for the return. Some muscle cars and trucks came with return-style mechanical pump fuel systems, but these are not the norm for older vehicles—most were dead-head systems, meaning that the fuel simply stops at the pump until it is sent on to the engine.

Many people assume you can use the 1999-and-up Corvette filter regulator with an LT engine, but this is not the case. While the engine will run, it will be starving the pump on the top end. This unit has two lines (an input and an output) on one side for the fuel tank, and one output on the other, which goes to the engine.

This preset regulator provides the correct 60 psi to the engine, which pressurizes the entire fuel line while pumping the excess fuel back to the tank. This is typically mounted as close to the tank as possible to minimize the

length of feed and return lines to the tank. This is not recommended for an LT engine swap. For a full dual-line system with the pump in the tank, a filter between the pump and the fuel rail is the preferred method if you are not running the factory PWM system (the filter is required for all fuel systems). It is best to filter the fuel as soon as possible, keeping the fuel lines clean.

Installing new lines is a fairly simple process, but it can be nerve racking at the same time. There are three ways to accomplish this task: run braided hose, bend new hard lines, or install pre-bent hard lines. Using pre-bent hard lines is the simplest method if the vehicle has the fuel tank in the stock location.

Pre-bent lines, such as those from Classic Tube and Tube Tech, are patterned after the original lines in the car and should fit just like the originals. That is not to say that there are not compromises and tweaks that must be made along the way.

Bending and installing custom lines most effectively transports fuel the length of the vehicle, but it is much easier said than done. This is a challenging task that requires some metalworking skills, and therefore the task will be frustrating at best for the novice. There are tubing makers, such as Classic Tube, that offer custom bending services using coat hangers or other wire. A pattern is bent by hand and sent to the maker. They, in turn, will bend a set of hard lines to your specifications and ship them to you. This ensures quality bends with proper flare where you want them—without kinks—and without the aggravation of doing it yourself.

The other option is to use flexible hose for the long runs. This works, but braided hose should be used rather than plain rubber hose to protect from road debris damage. The chance of road debris snagging a long, braided fuel line is much higher than with a hard line. Rubber lines are not the best option either. Rubber lines dry out and crack much faster than hard lines corrode, so you will have to replace the rubber lines eventually.

The Camaro has good hard lines up to the engine bay, so we chose to use them for the fuel feed. The larger line is 3/8 inch, which will work perfectly.

We used a tubing cutter and removed the flare because we will make our own.

We are using a tube sleeve and nut to create an AN flare for the pressure sensor. These are from Earl's Performance Plumbing and are -6 AN, which is the correct size for 3/8-inch line. The tube nut goes on first, then the sleeve, then flare the line.

AN fittings require a 37-degree single flare (never a double flare), which we made using this portable tool from Matco. The standard 45-degree flare commonly used for brake and non-AN fuel lines will leak with AN fittings.

Once flared, the sleeve slides up to the flare. The nut will thread onto the fitting, creating a tight seal.

We installed the fitting, which is an adapter port for the pressure sensor.

Phantom Pump System Installation

Until recently, installing an in-tank fuel pump meant buying a custom tank or doing a lot of dangerous cutting and welding on the tank. Thanks to some ingenuity from Aeromotive, that is no longer the case. The Phantom series of in-tank pumps allows you to quickly, easily, and safely install an in-tank pump into most stock fuel tanks.

For the 1971 Buick GS that we are LT swapping, we opted for this new pump system. The GM A-Body tank is a bit problematic when it comes to this install. There is only one good place to locate the pump, so you have to be careful when setting it up. These pump kits will work in any fuel tank that is between 6 and 11 inches deep.

The main issue with the Phantom system is that is needs a fairly flat surface to be installed. Most tanks are corrugated,

and Aeromotive knows this. You can install the system on uneven surfaces up to 1/4 inch deep by way of the included high-density rubber gasket. The A-Body tank has pretty deep ribs on it, so on 1969-and-older tanks, the front passenger's side is really about the only location you can use. Depending on the model of vehicle, there may or may not be vents at that location. The vent tube runs the length of the tank, so it has to be removed and plugged.

Before beginning this process, make sure that the tank is empty of fuel. It is a good idea to rinse it out with water and dry it out with some compressed air.

Once the area is picked, you have to cut. We used a 3¼-inch hole saw to make the hole. Go slow while cutting. Most of the metal shavings will stay outside the tank, but you

1 Each Phantom fuel system comes with baffle, pump, install ring, and all related hardware. They install easily into the A-Body tank.

2 I started by cleaning the tank and marking the location for the pump. There is really only one suitable spot, the front passenger's corner.

3 Drill the tank for the center of the mounting hole. I pre-drilled this because hole saws have a bad habit of breaking or jerking your hand when you go through the metal and the blade catches on the steel. Always practice safe drilling.

4 Cut the hole using the correct-size hole saw. A good quality bi-metal saw is necessary.

5 *Inside the tank is a long vent tube. Some A-Body tanks have this, some have just a short port. Remove the line if it is in the way.*

6 *To fill the hole, run a 1/4-20 tap through the hole.*

7 *Add some high-strength threadlocking compound to a stainless-steel bolt and thread it into the hole.*

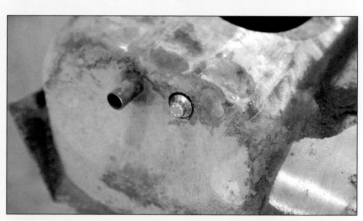

8 *Add a nut and washer on the inside of the tank to help seal the hole. It won't leak now.*

9 *Place the drill jig into the hole and align it so that the holes are not in the side of a rib (very important), and drill two opposing holes.*

10 *Place a single bolt into one of the two holes to secure the jig in place and keep it from twisting while drilling the remaining holes.*

Phantom Pump System Installation *continued*

will end up with some inside. That is okay because it should be vacuumed out later anyway. The kit comes with the necessary drill jig that fits perfectly inside the 3¼-inch hole.

Place the jig in the hole; make sure none of the drill holes run through the angled section of any corrugations in the tank. The mounting studs must be through flat metal. Drill two holes opposite each other, and then use the supplied bolts and nuts to secure the jig in the tank. Once secure, drill the rest of the holes and remove the jig. At this point, clean the tank with a vacuum to remove any metal shavings.

Now the foam baffle is cut to size. Measure the tank depth and add 1 inch. Cut the foam to that length. You want it a little longer than the tank to keep it in place. The kit comes with a stud ring that installs from inside of the tank. It is split on one side to go into the hole. Install this and pull the studs through the sheet metal. Drop the drill jig over the studs and loosely thread a couple of nuts onto the studs to hold it in place. Press the baffle through the hole and position the foam around the stud ring.

Using the tank depth measurement, assemble the pump and filter sock and set the max depth to the tank depth. Cut the pump mounting arm to accommodate this

11 *Measure the tank at the bottom and cut the foam baffle 1 inch longer than the total depth. A razor blade or serrated knife is the easiest way to cut the foam.*

12 *Place the baffle inside the tank. You must fold it onto itself, but it conforms relatively easily.*

13 *To secure the pump housing, place the split ring into the tank, align the studs, and press it up through the holes previously drilled.*

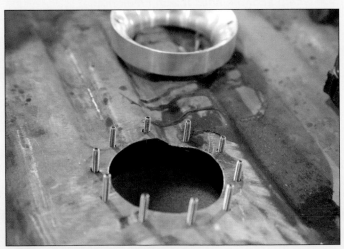

14 *The studs should be fully exposed. The foam baffle goes around the outside of the ring.*

15 *The pump gets a filter sock, and the rubber sleeve slides over the pump body.*

depth. It is critical that the pump sock sits on the bottom of the tank, otherwise you can starve the pump for fuel, which will kill it. Cut the supplied fuel hose to match the installed depth and assemble the pump to the housing using the clamps provided. Don't forget to install the wiring harness.

Slip the gasket into the pump assembly, remove the drill jig, slide the pump into the tank, and drop the housing over the studs on the ring. Thread the nuts and washers onto the studs, and tighten them in a crisscross pattern evenly to seat the housing and not distort or break the studs. It is possible to overtighten the studs and break them, so be careful. Hand tight is all you need; don't use an impact wrench.

At this point, the tank can be installed in the vehicle. For

A-Body cars, space the tank down about 1/2 inch to get the clearance for the pump assembly. This can be done with foam or wood blocks or by fabricating two metal wedges. Unfortunately, this is necessary for clearance. Another option is to cut or dimple the trunk pan over the housing. Some cars have really tight spacing above the tank, so spacing the tank down is a common requirement.

With the tank installed, wire it up and run the lines. This system is capable of return or returnless plumbing and can be PWM controlled for LT-series or aftermarket ECM control. The housing uses -6 fittings for ease of use with 3/8-inch hard fuel lines, making it perfect for factory or aftermarket fuel line. ∎

16 *Measure the tank depth again, this time for the pump depth.*

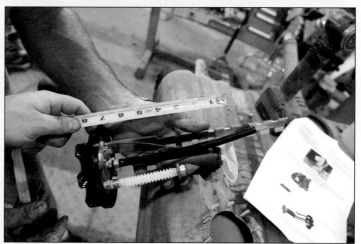

17 *Transfer that measurement to the pump housing and position the pump to touch the bottom of the tank. An A-Body tank should measure 7 inches, but it is critical that you measure it on your tank for absolute certainty.*

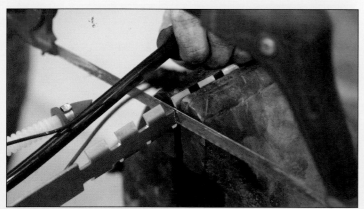

18 *Cut the rail a little short so that it does not impede the placement of the pump. A hacksaw makes quick work of the bracket. One notch short is suitable.*

19 *This pump is being used with the LT1 PWM system, so I drilled a 1/32-inch hole into the brass plug. This is a pressure vent and is only required for PWM returnless usage. This pump works great for return systems without this modification.*

Phantom Pump System Installation *continued*

20 Secure the pump to the housing and the fuel line to the top with band clamps. The wiring simply plugs in.

21 A foam pad goes onto the top of the studs and then the pump housing drops into the tank. Be careful not to pinch the wires.

22 Tighten each stud in a crisscross manner until the housing is sealed tight. These bolts do not require a lot of torque; don't over-torque or they will break. Hand-tightening with a 1/4-inch wrench is good. Connect the wiring. You can make a quick-connect with spade terminals.

23 Underneath the car, install a section of sound mat over the pump to absorb any noise from the pump.

24 The kit comes with some foam padding. The pump is too tall for the car, so I built up the front of the tank with three strips of foam in a pyramid shape. This places the tank low enough to clear the pump.

25 Before installing the tank, install the fuel lines. These short lines connect to the hard lines under the car. The stock tank straps work perfectly with 1-inch-longer bolts.

EXHAUST SYSTEMS

Unlike the older SBC exhaust systems, the factory components of a Gen V LT-series engine are actually pretty good. To squeeze out every available drop of fuel economy to meet the strict modern standards, the factory exhaust has to perform well, while being quiet and efficient for emissions too. This means that the stock manifolds are acceptable for swaps, and in many cases actually work better than any aftermarket offerings. There are a few reasons for this, but the bottom line is that getting good flow out of the Gen V engine is not hard.

The ECM has to monitor the exhaust system, which is critical to any EFI system, and even more so in a direct-injection engine.

Inside the combustion chamber, the fuel is injected at high pressures; any restriction of flow is a bad day for the engine. In stock trim, there are four oxygen sensors: two just behind the exhaust manifolds and two more just after the catalytic converter pipe, which is actually an assembly of two or three catalytic converters.

If you are planning on using catalytic converters for legal reasons (varies by state), then you will likely want to retain the rear oxygen sensors, but they can be deleted. If you are not using catalytic converters, then delete the rear oxygen sensors in the ECM programming. Most pre-programmed ECMs from harness companies already do this.

Check with local and state laws regarding engine swaps and emissions equipment. These laws vary by state, and when you get a vehicle inspection, it would not be good to be missing emissions equipment should it be required. Most states have classic car exemptions, so it all depends on the age of the vehicle.

Exhaust Manifolds

There are several factory manifolds for Gen V LT-series engines, including one for trucks, one for

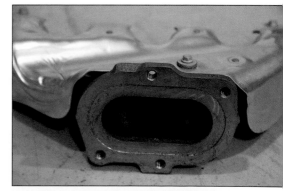

This flange is tricky to work with because you must find a lower flange that will match it, and those are currently not available in the aftermarket. So, you would have to find a factory pipe to scavenge.

LT1 exhaust manifolds are really functional for most swaps because they dump in the center of the engine and have a very large lower flange. This gets in the way of most chassis designs.

Corvettes, and one for the other passenger-car models. The following list provides the GM part numbers for the factory exhaust manifolds.

Corvette

The Corvette manifolds are center dump, which does not clear any aftermarket motor mount adapters.

Non-supercharged 6.2L (LT1)
Left: 1269724
Right: 12629725

LT4 supercharged engines
Left: 12629726
Right: 12629727

Camaro

The Camaro manifolds are center dump, which does not clear any aftermarket motor mount adapters.

Left: 12629728
Right: 12629729

Trucks

Manifolds are the same for 5.3L and 6.2L engines. These manifolds are swept back, so they dump to the rear of the engine.

They clear most aftermarket motor mount adapters and are suitable for many LT swaps. The reality is that there is just not much available in the aftermarket for LT-swap headers, so most swaps will use a version of the truck manifolds or headers.

Left: 12629337
Right: 12629338

If you are going to use factory headers in a Gen V LT swap, then you are going to need some flanges. The factory flanges are not readily available in the aftermarket, so you have to build your own. You could cut them off of a factory Y-pipe, which is also the catalytic converter for 2014-and-up GM trucks, but this might be a fairly costly option. The catalytic converters cost about $1,400 brand new, so the used ones will not be cheap.

Building flanges is not that complicated, but it does require attention to detail and some heavy steel plate. If you do not have the ability to fabricate them, any local machine shop or steel fabrication shop could make them quickly, and it should not

cost much more than a couple hundred bucks.

To make a set of flanges, you need the following:
- 1/4-inch plate steel
- Torch or plasma cutter
- Carbide drill bit set
- Drill
- Cardboard (for template)
- Pencil
- Grinder
- Center punch

The flanges begin as a piece of flat plate steel. The minimum thickness for the flanges is 1/4 inch, and 3/8 inch would be even better. You want a thick flange to avoid warping. We actually had an original flange that happened to come attached to one of the manifolds from the salvage yard, but we only had one. The yard that we bought our engine from did not have any others available, so we used the one we had as a template.

The factory manifold (in this case a 2015 Chevy truck manifold) can be used to trace the shape of the flange onto a piece of cardboard to create a template. We traced our flange several times to make extras because

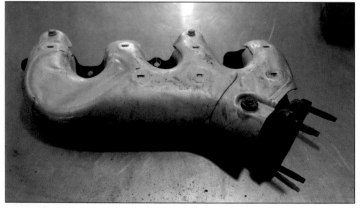

Trucks use a more common swept-back design that looks more like what we are used to seeing. These are actually pretty good manifolds, and they are more than suitable for most swaps.

Once the heat shields are removed, you can see that these are closer to shorty headers than block manifolds.

In the 1967 Camaro test chassis, the passenger's side fits great with the stock truck manifold.

Unfortunately, the driver's side does not clear the steering box. If this Camaro was converted to a rack and pinion, it would work great.

This set of Holley truck headers is closer to clearing. It might be possible to modify the number-1 primary tube and make them fit.

The manifold bolts need a liberal coating of anti-seize because the heads are aluminum and the bolts are steel. If you don't do this, the bolts will gall in the threads, and that is a great way to ruin a set of heads.

we are doing several swaps at the moment. If you make several, you can cherry pick the best ones to use for your project.

We used a plasma cutter to cut out each flange and then used the grinder to clean them up. A large metal-cutting band saw will produce cleaner results. The center hole for the exhaust itself can be cut with the plasma or with a hole saw; we used the plasma for this example.

The critical area for this project is the three bolt holes. If it is off

LT exhaust gaskets are laminated steel for a nice seal. They can be reused a few times if necessary.

even a little, it will not seal. Using the center punch, we marked the center and drilled a small 1/8-inch pilot hole in each flange bolt hole. Then, we used a series of increas-ingly larger drill bits to open up the hole. Do not do this in one step because 1/4-inch plate is hard enough to drill as it is; breaking a bit is no fun.

The flanges are pretty simple to make, and they work really well, but now there is another issue. Making the flanges means that you can't use the factory donut gasket. To rectify this problem, get some steel-lined exhaust gasket material from Fel-Pro, specifically Fel-Pro Pro-Ramic Gasket Material (part number 2449). This stuff is made from a sheet of perforated steel coated on both sides with ceramic gasket material. You can cut it with tin snips, and it will hold up to the heat of engine exhaust. If the flanges are flat, it won't have any issues with leaks.

Making a gasket is not difficult, and this process works on standard gasket materials as well. Non-steel-lined material is much easier to work with because you don't have to cut through the steel core. We made the gaskets with relative ease. Three tools are needed: tin snips, a hammer, and a set of hole punches. Our punches came in a set of six, and while they won't be used often, when they are needed, you will be glad you spent the $10 on them.

We used the new flanges as a pattern and traced the shape of the gasket with a pencil. This needs to be as accurate as possible. Next, we used a punch that was the closest to the required hole size and made the holes. A couple of blows with a hammer cut right through the gaskets without any tearing or jagged edges. Then, we used a pair of tin snips to carefully cut out the outside shape. Cut slowly and carefully. The metal core is sharp, so be careful and wear gloves.

The center hole is the toughest because it needs to be cut clean. We used the hole punch to get the center hole started and then used the snips to finish the job. Work slow and have lots of patience. It is a really good idea to make several sets of gaskets at once. That way, replacements are on hand if they need to be changed out. These will last for a while, but they are not going to last nearly as long as the factory-style donuts.

We caught a lucky break on our Camaro project when the factory GM truck manifolds actually fit. If they didn't, it would either mean buying a bunch of LT headers for stock cars and trying them or building our own, which we know from experience is difficult.

The driver-side manifold fits no problem, but the passenger's side is so close you might cry, because it doesn't fit as is. The outside bolt on the three-bolt flange runs right into the chassis. You could notch the chassis, but we wanted to see what it would take to leave the frame rail intact.

First, we tried sanding down the flange, but we needed to get

We were lucky and one of our L83 salvage engines came with one stock exhaust flange. We will use this to make our own. If you can find a pair, you can use them with the stock donut seal.

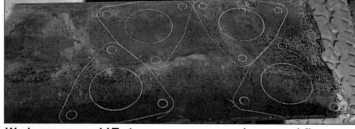

We have several LTs to swap, so we made several flanges. It is a good idea to make a few to have extras on hand just in case.

A plasma torch made quick work of the 1/4-inch plate steel we selected for the flanges. Work slow to have less cleanup to do later. Don't torch the bolt holes because those require precision drilling.

After they cooled down, we clamped each flange to the bench on a piece of scrap wood. Each hole was center punched for accuracy.

more drastic. We removed the stud, cut about half of the hole away, and used a Bridgeport mill to slot the hole inward about 1/2 inch. This can be done with a drill and some grinder work, so don't get scared. We will have to use a bolt and washers, but that is okay because it will seal.

As a side note, the Holley LT truck headers fit as well, but they have the same problem. A quick remedy is to cut the three-wing flange off and weld up a ball-and-socket flange—or even better yet, use a V-band flange. We wanted to show you how do it the hard way, so we stuck with the manifolds.

Now for the second part: you have to make your own exhaust flanges. It might be possible to use LS truck flanges, but we don't have any and they would need to be modified anyway, so we made our own with some 1/4-inch plate steel and a plasma cutter.

The driver's side fits great, but the upper bolt on the manifold was just a hair too long. We could cut it, but that would make it hard to remove later (the end is a hex for removal), so we notched

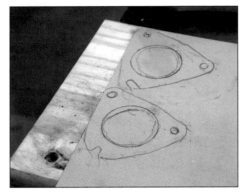

Because we can't use the factory donut gasket with our homemade flanges, we need to make our own gaskets. We traced the flange onto some Fel-Pro Pro-Ramic exhaust gasket sheet. This gasket is for the Camaro, which we had to modify to fit the chassis.

the upper hole in the flange so that it can twist into place. The big donut from the factory won't fit the flanges we made, so we also made our own gaskets. Fel-Pro Pro-Ramic exhaust gasket material did the trick.

The third-generation Camaro uses a single-pipe exhaust, so there isn't much room to run true duals. However, it can be done. We opted for a Magna-flow cat-back stainless system and made our own downpipes. The downpipes were built from 2½-inch mild steel tubing and some 3-inch stainless tubing from Summit Racing. They offer complete kits and single pipes, so getting the job done is easy. The trick is to order more than you think you need. They can always

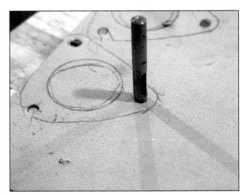

Using a hole punch, the bolt holes were cut away first.

We drilled each hole using three bits: a 1/8-inch pilot bit, an intermediate size, and the final 3/8 inch.

Then, we used some tin snips to cut the gaskets. There is a layer of steel mesh between the two layers of gasket material, so the snips are needed.

After punching a hole on the edge of the center opening, the snips were used to carefully trim away the center opening.

We made several sets of gaskets for future use. The nice thing about this is that the material is cheap and can make whatever you need.

Under the Camaro, the manifolds fit tightly to the chassis, and this is after we massaged the inner frame rail to clear. There is 1/4 inch of clearance, which should be enough.

Then, after a few blasts from some penetrating lubricant, the stud zipped right out.

On the driver's side, we had room for the stud, but to get the flange on, we notched the side of the lower flange at the upper bolt, which allows us to rotate the flange in place.

Because we are adapting this kit to the engine, we started at the back, laying out all of the pipes and muffler to the correct location.

The studs on the manifolds won't work with how tight the engine fits to the chassis, so we had to remove them. Some gentle heat was applied to loosen the rusty bolt in the manifold.

We needed to open the bolt location to get the necessary clearance to the chassis, which left an open notch for the third bolt.

Instead of building an entire exhaust from scratch, we opted for a Magnaflow exhaust kit for the 1987 Camaro. This gave us almost everything we needed to link up to the LT engine. The only thing we used extra of was a few bends of exhaust pipe that we ordered from Summit Racing.

This system uses wide band clamps with a seal better than the U-bolt-style clamps.

Once the exhaust was in place, we began working to connect the two pieces. Lots of measuring and marking of pipes occurs during this process to get the angles right.

The manifold flanges are first, and we tacked each lead pipe to the flanges. Only use a tack weld in case it has to be changed later.

A stand-up band saw is really helpful for this project. It is possible get by with a chop saw or portable band saw, but the cuts will not be perfectly square or on the exact angle that is needed, so more hand trimming is necessary.

We rotated and adjusted each bend until the entire line was completed. This is a single exhaust system because the torque arm gets in the way of running duals on this chassis. It can be done, but it hangs quite low.

This is the overall Y to the main exhaust line. Now we need to make an actual Y-pipe.

We used some scrap pipe for the exhaust along with some stainless steel V-Band clamps, which makes this exhaust removeable for servicing.

To make the Y-pipe, we notch a section into the main length.

We want a smooth-flowing exhaust, and it needs to be leak free. So, the notch was smoothed out using the edge of a belt sander.

Now we have a nice fish mouth, which is similar to how a roll cage is assembled.

Here is the actual Y-pipe. Is it perfect? Probably not, but it works. The angle of entry for the second line should be smooth enough to not cause turbulence.

We tacked all of the parts together under the car. The V-band clamps allow each section of the exhaust to be removed for servicing other components. This is critical, especially for a new engine swap, which will likely need a few tweaks before it is all finished.

The last part of the job is opening up the interior of the Y. We used a hole saw and long extension to reach the end of the pipe.

be used on the next car (or your buddy can use it.)

If your state requires vehicles to have catalytic converters, aftermarket converters can be used in the downpipes. The downpipes need to be separate from the rest of the cat-back system for service, so we used a pair of 3-inch V-band clamps at the Y-pipe to connect to the Magnaflow system. That was

made with two 3-inch 45-degree bends. The 2½-inch downpipes were connected with a pair of 2½- to 3-inch adapters that we had made at a local exhaust shop.

The beauty of the Magnaflow

system is that it uses the factory hangers, which we replaced with new, and provides the destination point for the downpipes. We installed the Magnaflow kit first, and then built the connector pipes. Plus, the Magnaflow kit has two beautiful polished stainless tips, so the Camaro will look and sound great.

Aftermarket Headers

There are very few options on aftermarket headers designed for engine swaps at the current time. A few are available from Hooker and BRP HotRods, but that is all there is at the moment. The issue is that General Motors not only changed the exhaust flange for the Gen V LT-series engines, but the angle of the exhaust flange itself is about 3 degrees sharper.

During the research process for this book, we worked with a few header manufacturers to create some Gen V headers by simply swapping the LT-series flanges onto a set of headers for the Buick GS built for this book. The result was a pair of headers where the collectors hit the bellhousing. This means that any manufacturer that wants to build LT-series swap headers has to completely reengineer the design, and that takes time. Eventually, headers will be available for the most popular models.

Hooker is already on the move with the recent release of the 520-70101352-R headers for first-generation Camaros. In fact, it has several full-length headers for the first-gen F-Body, including fitment for the factory subframe. It is set for the Detroit Speed subframe conversion. This is the first mass-produced set of LT-swap headers that are designed to work with the factory components.

BRP HotRods has several swap kits available for the Tri-Five Chevy, first-gen F-Body, 1964–1972 Chevy truck, 1964–1972 GM A-Body, 1978–1988 G-Body, 1970–1981 F-Body, 1973–1987 Chevy truck, 1971–1976 GM B-Body, and 1988–1998 GM truck. This represents a huge segment of the LT-swap market, but the headers for these applications appear to be designed specifically for the rest of the BRP LT-swap

Don't forget the oxygen sensors, which must be mounted one on each side within 18 inches of the manifolds.

We used a Rotabroach-style hole cutter to make a clean hole in the pipe. These bungs are large, so a hole saw is necessary.

Each bung was pushed into the hole and then welded up. Get a good hot weld to avoid pinhole leaks, which will burn out the oxygen sensor.

There are a few sets of aftermarket headers available for specific swaps, and as time goes on, there will be more. These headers from MuscleRods are specifically for use with the BRP HotRods swap kits; it is unclear whether they will work with other swap kits and adapters.

The MuscleRods headers definitely fit very well, and they are stainless, so they look really good too.

platform, so it is unclear whether or not the headers will fit with other swap adapters. In addition, they are quite expensive at $1,149, although they are very nice quality. Built from stainless steel, they are worth the price, but that might be out of reach for the average low-budget swap project.

Until the aftermarket catches up with the increasingly popular LT swap, there are options beyond the factory manifolds. Most aftermarket LT-powered truck headers are a mid-length-style header, where the collector dumps at the bellhousing, about midway between the oil pan flange and the head deck. This makes 2014-and-up GM truck headers the best bet for most applications.

That doesn't work on every chassis, but it will get you pretty close.

Catalytic Converters

For those who live in emission-controlled areas, chances are high that some sort of catalytic converter needs to be installed with a Gen V LT swap. Most of the mystery lies in the construction of a catalytic converter. The most important part is the ceramic matrix.

Shaped like a honeycomb, the matrix is made predominantly of a ceramic material called cordierite. The honeycomb is created through an extrusion process in which lengths of the honeycomb shape are squeezed

through a die and supported by computer-controlled jets of air that keep the honeycomb straight as it leaves the machine.

Once the ceramic honeycomb is fired and set, it receives a washcoat of various oxides combined with the precious metals that function as the actual catalyst. The washcoat is used because it evenly disperses the metals throughout all the pores of the matrix. The metals are generally mixed to best utilize their individual properties; most catalytic converters in the United States use some combination of platinum, palladium, and rhodium. Outside of the United States, copper has been tried, but it will form dioxin, a toxic substance with carcinogenic properties. In other places in the world, materials such as nickel, cerium in washcoat, and manganese in cordierite are used, but each has its disadvantages.

Originally developed throughout the 1930s and 1940s for industrial smokestacks, catalytic converter inventor Eugene Houdry began to develop a catalytic converter for automobiles in the 1950s. The first catalytic converters were mandated in 1975. These

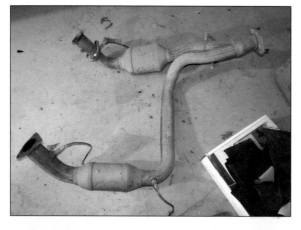

All LTs require catalytic converters as part of the stock exhaust. However, when swapping an engine, state laws determine whether or not they need to be kept. If they don't have to be kept, it is a lot easier and cheaper to build the exhaust. This is a stock Y-pipe with the dual cats and also the before and after oxygen sensors.

Inside the catalytic converter is a honeycomb of palladium, platinum, or rhodium. All of these are considered precious metals, which is why catalytic converters are expensive.

Magnaflow builds high-flow catalytic converters for performance applications. If you need to keep them, this is a really good way to go for adding them to your swap exhaust.

were a two-way type, which combined oxygen with the carbon monoxide and unburned hydrocarbons to form carbon dioxide and water. As the science progresses, more strict environmental regulations brought about a change to three-way converters in 1981. These more-advanced catalytic converters also reduce nitrogen oxide.

Catalytic converters use what is called a redox reaction. This means that once the catalyst is up to operating temperature (anywhere from 500 to 1,200°F), both an oxidation reaction and reduction reaction are occurring simultaneously. That sounds a little complicated, but it means that molecules are simultaneously losing and gaining electrons. These types of reactions are extremely common; photosynthesis and rust are both good examples of redox reactions.

In the first stage of the catalytic converter, the reduction stage, the goal is to remove the nitrous oxide and especially the nitric oxide. When nitric oxide is introduced to air, it quickly changes into nitrogen dioxide, which is very poisonous. The reduction stage works because the nitrogen molecule in the nitrogen oxides wants to bond much more strongly with the metals of the catalyst than it does with its oxygen molecules and the oxygen molecules would rather bond with each other, forming the type of oxygen that we breathe.

Once the oxygen molecules break off from their nitrogen molecules, the nitrogen molecules move along the surface of the catalyst, looking to make friends with another nitrogen molecule. Once it finds one, it bonds and becomes the stable, harmless nitrogen we find in our atmosphere. Once it becomes atmospheric nitrogen, its bond with the walls of the catalyst is weakened and the gas moves along to the second phase of the catalytic converter, which is oxidation.

Once the gases have finished in the reduction stage of the catalytic converter and we've eliminated all the nitrogen oxides, we are left with atmospheric nitrogen, atmospheric oxygen, carbon dioxide, carbon monoxide, water, and unburned fuel. The oxidation stage of the catalytic converter uses platinum and palladium, which want to bond with the various oxides.

The oxygen molecules bond with the surface of the catalyst and break up, eventually finding carbon monoxide molecules to bond with, creating carbon dioxide. The carbon dioxide bond is stronger than the bond with the catalyst and moves through the matrix, allowing the process to begin again. At the same time that this is happening, some of the free oxygen molecules begin to bond with the unburned fuel (hydrocarbons) and are changed into water and more carbon dioxide.

This brief history of the design of catalytic converters is important because it goes to show the advancements in catalytic converter technology. The original catalytic converters were very inefficient and clogged up quickly, causing serious performance issues. Modern performance catalytic converters are designed for free flow while doing their job of cleaning the exhaust gases. Magnaflow makes high-flow CARB-legal and New York-legal catalytic converters that ensure the emissions of your LT swap will match the necessary specs for your area.

Air Intake

While the exhaust gases are managed by the mufflers and exhaust system, the incoming air charge requires some attention as well. Unlike the exhaust, the intake system is very simple. An air cleaner element, some tubing, and the mass air flow (MAF) sensor are all that is required for the EFI Gen V LT-series engine.

It is possible on some swaps to simply install a cone-style air cleaner onto the throttle body with a built-in MAF sensor between the two, but more often than not, you need some sort of ductwork. General Motors specifies a 4-inch-diameter air intake tube for the Gen V LT-series engine, which can complicate the swap process, but these are relatively easy to fabricate.

The key to a good intake system is large, smooth bends in the piping. Air does not like to make abrupt turns, as this creates a vortex effect inside the tube that can drastically slow down the air. Slow air means less air in the engine. Short air filter elements require the air to make fast direction changes, and that will siphon off potential horsepower. The best bet is to make the intake tube as straight as possible, preferably grabbing the cooler air outside of the engine bay.

Advanced Clutch Technology
206 E. Ave. K-4
Lancaster, CA 93535
661-940-7555
advancedclutch.com

Aeromotive Inc.
7805 Barton St.
Lenexa, KS 66214
913-647-7300
aeromotiveinc.com

American Powertrain
2199 Summerfield Rd.
Cookeville, TN 38501
931-646-4836
americanpowertrain.com

AutoMeter
413 W. Elm St.
Sycamore, IL 60178
866-248-6356
autometer.com

Automotive Racing Products
(ARP)
1863 Eastman Ave.
Ventura, CA 93003
800-826-3045
arpfasteners.com

B&M Racing & Performance
9142 Independence Ave.
Chatsworth, CA 91311
818-882-6422
bmracing.com

BRP HotRods
5849 Rogers Rd.
Cumming, GA 30040
770-751-0687
BRPhotrods.com

Chevrolet Performance
chevroletperformance.com

Comp Cams
3406 Democrat Rd.
Memphis, TN 38118
Tech Support: 800-999-0853
compcams.com

Current Performance
6330 Pine Hill Rd., #16
Port Richey, FL 34668
727-844-7570
currentperformance.com

Dakota Digital
4510 W. 61st St. N.
Sioux Falls, SD 57107
800-593-4160
dakotadigital.com

DeatschWerks
415 E. Hill St.
Oklahoma City, OK 73105
800-419-6023
deatschwerks.com

Detroit Speed and Engineering
185 McKenzie Rd.
Mooresville, NC 28115
704-662-3272
detroitspeed.com

Dirty Dingo Motorsports
506 E. Juanita Ave., Suite 3
Mesa, AZ 85204
480-824-1968
dirtydingo.com

DynoTech Engineering
1731 Thorncroft
Troy, MI 48084-5302
800-633-5559
dynotechengineering.com

Earl's Performance Plumbing
19302 S. Laurel Park Rd.
Rancho Dominguez, CA 90220
310-609-1602
holley.com

Edelbrock
2700 California St.
Torrance, CA 90503
310-781-2222
edelbrock.com

EFI live
efilive.com

F.A.S.T.
3400 Democrat Rd.
Memphis, TN 38118
877-334-8355
fuelairspark.com

Flex-a-Lite
7213 45th St., Ct. E.
Fife, WA 98424
800-851-1510
Flex-a-Lite.com

Griffin Radiators
100 Hurricane Creek Rd.
Piedmont, SC 29673
800-722-3723
griffinrad.com

Holley
1801 Russellville Rd. (Zip: 42101)
P.O. Box 10360 (Zip: 42102)
Bowling Green, KY
270-782-2900
holley.com

Hooker
1801 Russellville Rd. (Zip: 42101)
P.O. Box 10360 (Zip: 42102)
Bowling Green, KY
270-782-2900
holley.com

HP Tuners
725 Hastings Ln.
Buffalo Grove, IL 60089
hptuners.com

ICT Billet
1107 S. West St.
Wichita, KS 67213
316-300-0833
ICTBillet.com

Jet Performance
17491 Apex Cir.
Huntington Beach, CA 92647
800-535-1161
jetchip.com

Koul Tools
928-854-6706
koultools.com

Lingenfelter Performance
Engineering
1557 Winchester Rd.
Decatur, IN 46733
260-724-2552
lingenfelter.com

Lokar Performance Products
865-966-2269
lokar.com

Magnaflow
22961 Arroyo Vista
Rancho Santa Margarita, CA
92688
800-824-8664
magnaflow.com

Miller Electric Manufacturing Co.
1635 W. Spencer St.
P.O. Box 1079
Appleton, WI 54912-1079
920-734-9821
millerwelds.com

Moroso
80 Carter Dr.
Guilford, CT 06437-2116
203-453-6571
moroso.com

MSD
Autotronic Controls Corporation
1350 Pullman Dr., Dock #14
El Paso, TX 79936
915-857-5200
msdignition.com

OPTIMA Batteries, Inc.
5757 N. Green Bay Ave.
Milwaukee, WI 53209
888-8OPTIMA
optimabatteries.com

Painless Performance
2501 Ludelle St.
Fort Worth, TX 76105
817-244-6212
painlessperformance.com

Pertronix Performance Products
440 E. Arrow Hwy.
San Dimas, CA 91773
909-599-5955
pertronix.com

PRW Industries, Inc.
1722 Illinois Ave.
Perris, CA 92571
888-377-9779
prw-usa.com

Red Dirt Rodz
4518 Braxton Ln.
Stillwater, OK 74074
405-880-5343
RedDirtRodz.com

Rock Valley Antique Auto Parts
800-344-1934
rockvalleyantiqueautoparts.com

Royal Purple, Inc.
One Royal Purple Ln.
Porter, TX 77365
888-382-6300
royalpurple.com

Speartech
3574 East State Rd., 236
Anderson, IN 46017
765-378-4908
Speartech.com

Summit Racing
P.O. Box 909
Akron, OH 44398-6177
1-800-230-3030
summitracing.com

Tanks Inc.
P.O. Box 400
Clearwater, MN 55320
320-558-6882
tanksinc.com

TCI
151 Industrial Dr.
Ashland, MS 38603
888-776-9824
tciauto.com

Trans-Dapt
12438 Putnam St.
Whittier, CA 90602
562-921-0404
hedman.com

VaporWorx
Newbury Park, CA 91320
805-390-6423
vaporworx.com